Climate Change and Public Health

Helmut Brand • Neethi Rao
Dhanasree Jayaram
Editors

Climate Change and Public Health

Governance and Challenges

Springer

MANIPAL
UNIVERSAL PRESS

Editors
Helmut Brand
Prasanna School of Public Health
Manipal Academy of Higher
Education (MAHE)
Manipal, Karnataka, India

Neethi Rao
Health and Human Development Team
Centre for Social and Economic Progress
New Delhi, Delhi, India

Dhanasree Jayaram
Department of Geopolitics and International
Relations, Manipal Institute of Social
Sciences Humanities and Arts
Manipal Academy of Higher
Education (MAHE)
Manipal, Karnataka, India

Jointly published with Manipal Universal Press
Copyright page of International Edition: This edition is not for sale in India. Customers from India should order the print book from Manipal Universal Press (MUP).
ISBN of the India edition: 978-81-971369-0-0.

Copyright page of India Edition: This edition is not for sale outside India. Customers outside India should order the print and electronic book from Springer Nature.
ISBN of the International edition (outside India): 978-981-95-0894-5

Co-publishing partnership between Manipal Universal Press (MUP) and Springer Nature Singapore Pte Ltd. The registered company address is 152 Beach Road, #21-01/04 Gateway East, Singapore, 189721, Singapore

ISBN 978-981-95-0894-5 ISBN 978-981-95-0895-2 (eBook)
https://doi.org/10.1007/978-981-95-0895-2

This Springer imprint is published by the registered company Springer Nature Singapore Pte Ltd.
The registered company address is: 152 Beach Road, #21-01/04 Gateway East, Singapore 189721, Singapore

Foreword

I am very honoured to contribute a foreword to this outstanding volume, titled *Climate Change and Public Health: Governance and Challenges*. This is very meaningful to me on a personal level given my decades-long friendship with DA Prasanna and with Prof Rajani Prasanna. It is very rewarding to see how their dedicated efforts and intense commitment over many years to education, science, technology, and public health have led to this excellent volume on one of the most critical issues of our time. I am also very excited to see this volume's effort to link the health aspects of climate change to governance issues, which has been a major focus of both my and Prasanna's professional lives. Although it skilfully summarizes a great deal of existing knowledge, it is focused on future challenges and potential solutions, a positive approach that DA Prasanna and Rajani Prasanna have always emphasized.

This book is a very well-edited series of chapters based on the 'Professor Rajani Prasanna Endowment Lecture Series on Climate Change and its Impact on Public Health', which was supported by a generous endowment from Prof Rajani Prasanna. She is a renowned art historian, academic, and philanthropist, with a broad interest in societal transformation and improving the human condition; I am thrilled to see her support extend to the critical climate issues that are the focus of this volume. DA Prasanna, her husband, is the Chief Patron of the Prasanna School of Public Health, Manipal Academy of Higher Education, Manipal, India, which has taken the initiative in preparing this volume. He is a pioneer in the MedTech sector in India and the founding CEO of Wipro GE, who has innovated new business models in making modern medical technology accessible and affordable while leading GE Health Care to a leading position in India. I am so enthused to see his remarkable leadership skills now being applied in the academic sphere, and this book is also a reflection of his vision to focus on perhaps the most critical public health issue confronting us today, and one that India must address effectively.

I was privileged to moderate the opening program of the Prof Rajani Prasanna Endowment Lecture Series which included presentations by Dr Henry Falk, Dr Poornima Prabhakaran, and Dr Nadav Davidovitch; all three are included in this volume. Several key themes were already clear at that time: linking public health and governance, highlighting the most thorough and timely data on the impact of climate change in India, and including leading climate change experts from outside India in the discussion.

The book is broadly divided into three sections. The first section begins by laying out the scale of the climate change problem, based on available evidence from authoritative sources such as the Intergovernmental Panel on Climate Change (IPCC) and others. The second section is a deep dive into specific public health impacts of climate change that are particularly relevant to the Indian sub-continent. This section includes chapters on air pollution, heat stress, coastal ecosystems, and freshwater systems. The third section includes chapters discussing the tools, approaches, and methods available to help policy developers and decision-makers address the challenges of climate change and its public health impacts. This last section also includes a chapter on how cultural heritage can serve as a tool for climate resilience in an Indian context. I would like to extend my congratulations to the many outstanding authors, editors, and staff who have contributed in such a significant way to the success of this volume and add a heartfelt thanks to DA Prasanna and Prof Rajani Prasanna for their leadership in sponsoring the Endowed Lecture Series and this important volume.

In summary, this book examines both the current and anticipated impacts of climate change on public health while emphasizing the long-term governance implications, opportunities, and challenges for managing these impacts. This book will be of great value to a broad audience of interested members of the public, scholars, early-career researchers, and academics, as it presents both the urgency of the climate change issues as well as potential ways to move forward to address the public health impacts of climate change.

Jagdish N. Sheth

Recipient of Padma Bhushan, 2020
Charles H Kellstadt Professor of Business
Goizueta Business School
Emory University
Atlanta, GA, USA

Genesis of Endowment on Climate Change

Why is climate change happening, and how does it affect public health? As is evident, climate change is a global crisis driven by human forces. The emission of heat-trapping gases primarily caused by burning fossil fuels has resulted in a cascade of problems, not limited to intense droughts, storms, heat waves, rising sea levels, deforestation, and warming oceans.

The Prasanna School of Public Health (PSPH) was set up at Manipal Academy of Higher Education (formerly Manipal University) in 2017 to address some of these issues. At the inaugural lecture, Dr Arun Kumar spoke of Chikungunya and Dengue epidemics in coastal India and how virology and public health research had brought them under control.

On a larger scale, during 2020–21, the fast-spreading COVID-19 pandemic affected the entire world like never before. Healthcare infrastructure and accessibility have become the most crucial national and global issues. India developed and produced effective vaccines locally. They were made available to all citizens, mostly free, by the Indian Government. A portion of these Indian vaccines were exported internationally to developing countries. This was the most successful India-made global solution for dealing with the deadly virus. The COVID-19 epidemic brought home nations' public health and health security as a top subject. India's successful fight against COVID-19 has built worldwide confidence in its ability to resolve the impending climate change disasters.

Amidst rising awareness of the effects of climate change in India, PSPH started a discussion series on 'Climate Change and Its Impacts on Public Health' by allocating a part of its research budget to this initiative. Sensing the need for a more extensive dialogue, I decided to fund the discussion series. It aimed to engage global thought leaders on how climate change impacts society and what measures they think communities need to take to mitigate the impact. PSPH has made a start.

Climate change continues to affect daily life, including our food, health, business, and politics. I hope the PSPH perseveres in its efforts to share ideas, and solutions, and offer recommendations that spark global dialogue on climate change-induced disasters. I assure you of my full support for this cause and wish it success in all its endeavours.

Rajani Prasanna

Introduction

This volume, titled *Climate Change and Public Health: Governance and Challenges*, aims to provide an overview of the multidimensional ways in which climate change and public health are interlinked. An underlying thread is the governance aspects associated with managing the health impacts of climate change at the global, national, and state levels and the interconnections between the two.

The book is an outcome of the 'Professor Rajani Prasanna Endowment Lecture series on Climate Change and its impact on Public Health' (https://www.youtube.com/channel/@ccdaph). This initiative is supported by a generous endowment from Prof Rajani Prasanna, a renowned art historian, academician, and philanthropist with an extensive interest in impactful research in art and societal transformation. Prof Rajani is also the spouse of DA Prasanna, Chief Patron, PSPH, MAHE, Manipal. DA Prasanna is a pioneer in the MedTech sector and the founding CEO of Wipro GE. Prasanna innovated new business models, making modern medical technology accessible and affordable and leading GE Healthcare to the number one position in India in the process.

The book addresses a multitude of issues on climate change with a focus on its interlinkages to public health, the geopolitical and cultural aspects, climate financing, policymaking, and specific manifestations of climate change. The book also covers specific topics such as air pollution and heat health that are especially relevant in the context of the Indian subcontinent and are explored in detail to synthesize the latest available evidence on these issues. One of the objectives of the book is also to explore different approaches and methodologies involving studying the interlinkages between climate change and health, including assessing the health impacts of climate change and designing policies to address climate and health governance. It covers both global and Indian contexts, thereby providing diverse perspectives on climate change and public health and striving to explore effective and inclusive ways of tackling this major challenge.

The book is edited by Prof Dr Helmut Brand, Founding Director of the PSPH, along with Dr Neethi V Rao, Fellow at the Centre for Social and Economic Progress, Delhi and Adjunct Faculty at Institute of Public Health Bengaluru; Dr Dhanasree Jayaram, Assistant Professor (Senior Scale), MAHE, Dr Vani Lakshmi R (Assistant Professor, PSPH), Dr Rahul Sheshan Clare (Coordinator and Senior Research Fellow, PSPH); and Alyssa Selma Martins (Sub-Editor).

The chapters are authored by eminent experts in their respective fields, including Prof Dr Henry Falk (Adjunct Professor of Environmental Health at the Rollins School of Public Health, Emory University, USA), Dr Atul Kotwal (Executive Director, National Health Systems Resource Centre, Government of India), Prof Dr M D Nalapat (Professor of Geopolitics and UNESCO Peace Chair at MAHE), Dr Liz Green (Program Director for Health Impact Assessment at Public Health Wales, UK), Prof Nadav Davidovitch (Professor at the Negev's School of Public Health, Ben-Gurion University), and Dr Poornima Prabhakaran (Director, Centre for Health Analytics Research and Trends, Trivedi School of Biosciences, Ashoka University), among others.

The book is broadly divided into three sections. The first section begins by laying out the scale of the climate change problem, based on available evidence from authoritative sources such as the Intergovernmental Panel on Climate Change (IPCC) and others. Moving from global to local, chapters in this section lay out the various health impacts of climate change and the policy actions that are being undertaken to address these impacts at various levels. The next two chapters of the book are a deep dive into specific public health impacts of climate change that are especially relevant in the context of the Indian subcontinent. This section provides a comprehensive overview of climate governance in India through the examination of the National as well as state-level measures. Respective chapters investigate the role of current programs such as the National Program for Climate Change and Human Health (NPCCHH) among other measures in shaping India's policies towards ensuring 'green healthcare facilities'. Through the analysis of the country's climate footprint and climate-health relationship, suggestions are put forth to ensure India's progress to climate-resilient health infrastructure. The next chapter outlines the association between both concepts through the lens of geopolitics. It discusses the approaches from an international perspective and narrows it down to a national standpoint.

This second section includes six chapters on particulate and air pollution, heat stress, coastal ecosystems, and freshwater systems. Each of the aspects is scrutinized to sketch the diverse ecosystems ranging from the Swiss alpine station of Jungfraujoch to the Western Ghats in India and their respective challenges and predictions. This section also discusses the tools, approaches, and methods available to help policy/decision-makers address the challenge of climate change and its public health impacts.

The last section reviews the upcoming trends in climate change and the health domain. The complexity of climate finance and the role of public finance, its players, and its dimensions are examined by the authors. It acts as a double-edged sword as there is a need to finance the innovations in climate change as well as its potential to affect the economy. They argue that there is a need for an organized and coordinated system between stakeholders to meet the financial needs of affected countries. In the final chapter, the authors assert that culture transcends aspects related to climate change, public health, governance, and climate justice. It focuses on how cultural heritage can serve as a tool for climate resilience in an Indian context.

The book does not seek to be a compilation of original research. Rather it draws from existing evidence and approaches to inform and stimulate thinking about contextual measures to address the health impacts of climate change. Thus, the overall book examines both the current and anticipated impacts of climate change on public health, while keeping an eye on the long-term governance implications, opportunities, and challenges for managing these impacts. Keeping in mind a broad readership of interested members of the public, scholars, early-career researchers, and academics, the book presents both the urgency of the issue and potential ways to move forward to address the public health impacts of climate change.

Manipal, India Helmut Brand

New Delhi, India Neethi V. Rao

Manipal, India Dhanasree Jayaram

Manipal, India Alyssa Martins

Contents

Editors and Contributors

About the Editors

Helmut Brand is the Founding Director of the PSPH, MAHE, and a Jean Monnet Professor of European Public Health and Head of the Department of International Health at Maastricht University. He is an advisor to the European Advisory Committee on Health Research of WHO Europe and the Expert Panel on "Investing in Health" for the European Commission. His expertise includes Public Health Medicine and European Integration in Health. He is a past president of the Association of Schools of Public Health in the European Region (ASPHER) and currently serves as the president of the European Health Forum Gastein (EHFG). He studied Medicine in Düsseldorf and Zürich and holds a Master's in Community Medicine from the London School of Hygiene and Tropical Medicine and the London School of Economics.

Neethi V. Rao is currently Fellow at the Centre for Social and Economic Progress where she anchors The Asian Collective for Health Systems working across ten countries in South and Southeast Asia to strengthen the Asian voice in global health governance. She is also Adjunct Faculty at the Institute of Public Health Bengaluru and a researcher specialising in intersectoral policy and governance, at the intersection of health and environment. Dr. Rao worked with the World Health Organization to support research and advocacy towards a successful World Health Assembly Resolution on Social Participation. With a Ph.D in biology from the University of Virginia, USA followed by over a decade of health policy and systems practice, Dr. Rao brings an interdisciplinary and systems-oriented lens to strengthening institutions for health in Asia and beyond.

Dhanasree Jayaram is an Assistant Professor (Senior Scale) at the Department of Geopolitics and International Relations, Manipal Institute of Social Sciences, Humanities and Arts, and Co-coordinator at the Centre for Climate Studies, MAHE, Karnataka, India. She was a Research Fellow at Centre Marc Bloch (CMB) and Guest Researcher at Freie Universität Berlin—under the Alexander von Humboldt Foundation's International Climate Protection Fellowship during 2022–23. She is also a research fellow of Earth System Governance, a member of the Climate

Security Expert Network, and a non-resident fellow of Finnish Institute of International Affairs. She holds a PhD in Geopolitics and International Relations from MAHE. Her primary fields of interest include global environmental politics, climate diplomacy, environmental security, gender and climate governance, and environmental peacebuilding.

She received the Next Generation Foresight Practitioners (NGFP)—Security and Technology award in 2022 to work on the geopolitical and security implications of solar geoengineering for the Global South. She pursued a visiting fellowship (Erasmus Mundus—short-term PhD) at Leiden University, the Netherlands, from 2014 to 2015, and a postdoctoral fellowship at the University of Lausanne, Switzerland, under the Swiss Government Excellence Scholarship during 2018–19.

About the Contributors

Kiran Kumar Bhatt is a research fellow at the Centre for Health Diplomacy, Department of Global Health Governance, PSPH, MAHE. His research areas of focus have primarily been on political economy, international financial institutions, public health financing, health diplomacy, and economic warfare. Before his current role, he worked as a political intelligence analyst. He has an MA degree in Geopolitics and International Relations from MAHE. His undergraduate degree was in Economics (Honors) from Christ (Deemed to be University).

Benjamin Brem is a scientist in the Laboratory of Atmospheric Chemistry at the Paul Scherrer Institute (PSI) in Switzerland. His expertise includes the long-term monitoring of the physical and chemical properties of atmospheric aerosol particles at the high-altitude station Jungfraujoch. Before this appointment, he was working at Empa and was involved in the establishment of the first global particle emission standard for civil aviation gas turbines. Benjamin holds a PhD in Civil and Environmental Engineering from the University of Illinois.

Rahul Sheshan Clare serves as the coordinator for both the Centre for Digital Health and Technologies and the Innovation Cell, MAHE. With over three years of clinical experience as a trained doctor in Udupi, Dr Rahul also holds a master's degree in Advanced Health Services Management from Griffith University, Australia. He served as a Health Technology Assessment Project Officer at Queensland Health, Australia, before joining MAHE as Executive Assistant to the Founding Director (PSPH) in 2018. His research interests include health insurance, digital health, entrepreneurship, hospital administration, healthcare management, project management, quality management (ISO 9001: 2015 lead auditor certified by CQI/IRCA), and health technology assessment.

Lubna Dada is a scientist at the Laboratory of Atmospheric Chemistry at the Paul Scherrer Institute. Her expertise includes molecular understanding of gaseous precursors responsible for new particle formation from field observations and chamber measurements. She also has an interest in extreme events carrying pollutants to the central Arctic. Before this appointment, she was a member of the steering committee of the Aerosol and Haze Project at the Beijing University of Chemical Technology responsible for quantifying the contribution of gases and newly formed particles to Beijing's haze. Dr Dada holds a PhD in Natural Sciences from the Institute for Atmospheric and Earth System Research/Physics at the University of Helsinki.

Kaspar R. Daellenbach is a senior scientist at the Laboratory of Atmospheric Chemistry at the Paul Scherrer Institute. His expertise includes understanding the sources and formation of particulate matter in various environments ranging from pristine environments to highly populated megacities. His research focuses on the impact of particulate matter on human health, to improve the overall understanding of the role of anthropogenic and natural emissions in the process today and in the future. Earlier, he was at the University of Helsinki studying haze formation in Beijing funded by the Swiss National Science Foundation. Dr Daellenbach holds a PhD from the Swiss Federal Institute of Technology in Zurich.

Nadav Davidovitch, MD, MPH, PhD is an epidemiologist and public health physician. He is a Full Professor and Director of the School of Public Health at the Faculty of Health Sciences, Ben-Gurion University of the Negev in Israel. He has served as Chair of the Department of Health Systems Management at the University and the Centre for Health Policy Research in the Negev. His research interests include health policy, public health, vaccination policy, one health/eco-health, comparative healthcare systems, health and mobility, public health ethics, health inequities, and global health.Prof Dr Davidovitch serves on several international and national committees, among them: the Governing Board, European Public Health Association; Association of Schools of Public Health in the European Region (ASPHER) COVID-19 Task Force; Israel advisory committee for COVID-19. He authored or co-authored over 160 papers and book chapters, coedited 6 volumes and books, and published his work in leading medical and health policy journals, such as the *New England Journal of Medicine*, *Lancet*, *Emerging Infectious Diseases*, *Vaccine*, *Social Science, and Medicine*, and *Law and Contemporary Problems*. Since the outbreak of COVID-19 in Israel, he has been involved in research and the formulation of health policy and has advised various agencies in Israel and abroad on the need to make structural changes in the health system, with an emphasis on social issues and addressing health gaps.

Imad El Haddad is the Head of the environmental Molecular Science Group in the Laboratory of Atmospheric Chemistry at the Paul Scherrer Institute in Switzerland. His work is focused on urban emissions and their impacts on the atmospheric composition, human health, and the Earth's climate. He developed multiple analytical and numerical techniques to characterize the aerosol composition and sources in environments ranging from the Arctic to polluted China. He has refined the understanding of atmospheric aging of complex emissions through smog chamber experiments. Simple parameterizations of the resulting data have been implemented in regional and global models to determine the impact of aerosol pollution on our climate and health.

Henry Falk, MD, MPH is a pediatrician who has focused primarily on public health. He is currently an Adjunct Professor of Environmental Health at the Rollins School of Public Health, Emory University. Formerly, he was Director of the National Centre for Environmental Health at the Centre for Disease Control and Prevention (CDC) and Director of the Agency for Toxic Substances and Disease Registry (ATSDR), both in Atlanta, GA and part of the Department of Health and Human Services (DHHS), USG. He is a retired Rear Admiral/Assistant Surgeon General of the US Public Health Service.

Taruna Juneja Gandhi is a public health professional with more than 11 years of experience in policy and program development, management of large-scale public health projects, health research, capacity building, and technical documentation. She is an External Assessor for public health facilities under National Quality Assurance Standards (NQAS) and has Six Sigma Certification from the Indian Statistical Institute, Mumbai. Having worked in private, public, and development sectors, she has extensive experience and knowledge of public health scenarios in India. Dr Taruna has worked on various projects with the Ministry of Health and Family Welfare (MoHFW), National Health System Resource Centre (NHSRC), Public Health Foundation of India (PHFI), the WHO—India and SEARO, International SOS, and other consulting firms.

Liz Green is a Consultant in Public Health/Program Director for HIA at the WHO Collaborating Centre (WHO CC) on "Investment for Health and Wellbeing" in Public Health Wales. She has extensive knowledge and practical application of HIA, "Health in All Policies," and has worked on many HIAs of varying strategic levels, complexity, and topics including the comprehensive "Health Impact Assessment of Climate Change in Wales." She is a Visiting Professor at the WHO CC for "Healthy Urban Environments" at the University of West of England, UK, and has been recognized as a Woman of Influence in 2022 for her work in Spatial Planning and has published extensively on HIA.

Nitya Mohan Khemka is the Director of Global Alliances at PATH, a global Health think-tank focusing on health equity. She is also the Director of Policy at the N J Khemka Foundation. Nitya is a Visiting Fellow at Judge Business School at the University of Cambridge where she researches topics spanning gender inequality, poverty, and human development and lectures on sustainable development and gender. A Fellow Commoner for Clare Hall College, Nitya advises the college on its academic programs and fundraising strategy. She serves on the Women's Leadership Board of Harvard's Kennedy School of Government, supporting the Women and Public Policy Program.

Uday Khemka is Vice-Chairman of the N J Khemka Foundation (oldest Indian foundation focused on climate change mitigation since 1992) and the SUN Group (diversified, including renewable energy, and electric mobility). He Co-chaired the Clean Energy Finance Forum established by P M Modi and President Obama to catalyze investment towards India's clean energy targets. The key long-term focus is institutional climate finance, renewable energy, and electric mobility. Currently, he is Senior Climate Advisor to the Long-Term Investment Organization, representing 45 of the world's largest sovereign wealth funds (over $13 trillion in AUM). Advisory boards: The Climate Group India, Oxford Martin School, Indian School of Business, etc.

Atul Kotwal is currently serving as Executive Director of the National Health Systems Resource Centre (NHSRC), an apex body for technical assistance under NHM, MoHFW, GoI. With more than 38 years of experience, Dr Kotwal is a medical professional, public health researcher, administrator, educator, mentor, and teacher with significant contributions to the field of Public Health. He has diverse experience including serving in the armed forces, as a Public Health Advisor in Botswana, Africa, and providing technical support to GoI in the erstwhile Planning Commission. His areas of expertise include health systems, policy and planning, primary healthcare, urban health, health technology, epidemiology, implementation, and interdisciplinary research. Dr Kotwal has received numerous awards for his contribution to the Armed Forces, the Sena Medal (Gallantry), and Vishisht Sewa Medal (VSM), being the notable ones. With more than 150 publications in indexed journals, his contribution to research, evidence generation, knowledge sharing, and providing academic support in the field of public health is invaluable.

Satheesh Kumar is working as an Adjunct Professor at Manipal Academy of Higher Education, India. He has been involved in research on the seasonal and annual variations of greenhouse gases to retrieve the radiative forcing at a coastal location in India. Owing to his pioneering contributions in the chemistry and transport of trace gases that induce climate change, the Committee on International Space Research (COSPAR) selected him with an executive membership. He is a member of the Indian Aerosol Science and Technology and a fellow of the Earth Science Society of India. He has published more than 65 publications in international journals.

Alyssa Selma Martins is a sub-editor at PSPH. Her interest areas include migration (trends and policies in the local context and the European Union), climate (security challenges and economic impacts), health diplomacy, political dynamics of identity, human rights, and culture studies. She holds a master's degree in European studies from the MAHE. Her specializations include culture, literature, and politics.

Juliane Mirow is an MSc Global Challenges for Sustainability student at the European University Alliance CHARM-EU, currently studying at the University of

Barcelona. Her academic expertise includes Planetary Health and Health System Decarbonization. Besides her studies, she is involved in multiple public health initiatives focusing on the climate change and health nexus. Juliane Mirow holds a BSc degree in European Public Health from Maastricht University.

Joy Merwin Monteiro is an Assistant Professor at the Department of Earth and Climate Science at the Indian Institute of Science Education and Research Pune. His research interests lie in the field of extreme heat in South Asia, especially its meteorology, modelling, and impacts on health. Joy holds a PhD in Atmospheric Science from the Indian Institute of Science, Bengaluru.

M. D. Nalapat was appointed India's first Professor of Geopolitics and UNESCO Peace Chair by Manipal University in 1999, a position he still holds. An Executive Committee Member of the Editors Guild of India, a Member of the Indian School of Social and Economic Sciences, and an Associate Member of the National Institute of Advanced Studies and the United Services Institution, Prof Nalapat is also the Editorial Director of ITV Media Network, having previously edited the Mathrubhumi and the Times of India. He has lectured extensively in India, the US, the UK, Austria, China, Taiwan, and other countries and has originated several concepts, including that of the containment of China, Asian NATO, and Southern Asia rather than only South Asia as India's hinterland and the concept of the proxy nuclear state. He writes extensively in national and international publications and has authored nine books.

Sanjay Pattanshetty is a Professor and Head of the Department of Global Health Governance and Coordinator of the Centre for Health Diplomacy at PSPH, MAHE, Manipal, Karnataka, India. He has over a decade of experience in Public Health policy education program development, implementation, field research, and practice. He has several scientific projects and publications in reputed journals and has contributed to policy briefs in relevant areas. He completed a Doctor of Medicine in Community Medicine (MD) from MAHE, and a double master's in public policy and human development with a specialization in Foreign Policy and Development from United Nations University and Maastricht University, The Netherlands.

Poornima Prabhakaran is the Director of the Centre for Health Analytics Research and Trends, Trivedi School of Biosciences, and a Senior Research Scientist at the Centre for Chronic Disease Control. She is a physician and epidemiologist by training with a research focus on a life-course approach for chronic diseases and the impacts of environmental risk factors on health. She was previously Director of the Centre for Environmental Health at the Public Health Foundation of India. Besides leading research and capacity-building for environmental health research, Poornima has worked with the Ministry of Health and Family Welfare's National Program for Climate Change and Human Health providing an implementation framework for national and sub-national governments to develop climate-resilient and environmentally friendly healthcare facilities.

Andre S. H. Prevot is the research group head at the Paul Scherrer Institute and an adjunct professor at ETH Zürich in Switzerland. His expertise includes advanced source apportionment of atmospheric particles in Europe and Asia using the latest mass spectrometric and other methods in the field and the laboratory. He is also performing emission/aging studies of various emission sources focusing on the organic components of the atmosphere. He has published more than 450 peer-reviewed articles and is a highly cited researcher (Clarivate) every year since 2014.

Deepta Sateesh is a design researcher, educator, architect, and planner, working in contentious landscapes. Her research practice is focused on creating new pathways in design, education, and environmental policy. Her research in the Western Ghats of India gathers wet ontologies, situated practices, movement, and the politics of the colonial eye, and draws from design, environmental humanities, and philosophy. She is also a dancer, wanderer, and a photographer. Deepta is the Director of the Design + Environment + Law Laboratory and Odde Research Centre, and Dean of Research and Collaborations, at Srishti Manipal Institute of Art, Design and Technology, MAHE.

Sarang Satheesh is working as an HVDC Control and Protection Systems Engineer at General Electric in the United Kingdom. His research focuses on the development and implementation of control systems and protection strategies for Power network grids which are integrated into green technology. Sarang received his master's in Electrical Engineering from the University of Nottingham for his unique contribution to the development of a Drive cycle-based design of electric machines for hyper cars. The university selected him as International Student Ambassador to illuminate undergraduate students to the global technical mission of the university. He is an active member of IEEE and Young Professional Ambassador of IET, Northwest Midlands, UK.

Hardik Shah is a PhD student at the Department of Earth and Climate Science at the Indian Institute of Science Education and Research, Pune. His research interests lie in the field of atmospheric dynamics and its application to heat and moisture transport in South Asia.

Abhiyant Tiwari is the lead climate resilience and health consultant at NRDC India. His interests include environmental, climate, and disaster risks to public health, and he works to advance evidence-based adaptation and mitigation policies and actions in these areas. He has extensive experience in working with academia, governments, and non-profits on public health research, program and policy development, management, and evaluation. Right from the development and implementation of the first Heat Action Plan (HAP) of South Asia in the city of Ahmedabad in 2013, as a public health researcher and practitioner at the Public Health Foundation of India, Abhiyant has worked with knowledge partners like Natural Resources Defense Council and policy partners like national and sub-national governments on developing, implementing, and scaling heat wave health.He is also a member of the

Management Committee of Global Heat Health Information Network: a WHO-WMO Joint Initiative, Regional Technical Working Group for Safe and Disaster Resilient Hospitals in PEER - South Asia, Indian Meteorological Society, and a fellow of LEAD India, and Climate Reality. He is trained as a Master of Public Health with a specialization in Environmental Health Sciences from Harvard T H Chan School of Public Health.

R. Vani Lakshmi is an Assistant Professor at the Department of Data Science. She holds a master's degree in Statistics and Management and a doctoral degree in Statistics from Pondicherry University (A Central University) in India. She also serves as the Editorial and Social Media Coordinator of the PSPH. She has taught courses in the MSc Data Science, MSc Biostatistics, and MSc Digital Epidemiology programs offered at MAHE since 2019. Her areas of academic interest include quality management (ISO 9001/14001 and Lean Six Sigma), applied statistics, and machine learning apart from creative and technical writing.

Julien Venne is a seasoned expert in innovation, design, and social sciences with a passion for social impact, health and wellbeing as well as democratic and environmental challenges. His company MalsamaMind develops digital solutions and engagement services that generate a positive impact on People and the Earth. At the academic level, he is the Head of the Department of Social and Health Innovation at PSPH, MAHE, and is also affiliated with the United Nations University (UNU-MERIT). He is a member of the WHO initiative "ATACH" related to the development of low-carbon and resilient health systems. He is also a member of the Planetary Health Alliance (PHA). His research activities are related to Digital Transformation, Data and AI Governance and Trustworthiness, Digital Health and Wellbeing, Public Health, Planetary Health, Digital Democracy, Behavioral sciences, and Post-growth Economics.

Climate Change and Public Health: Interconnections and Challenges

1

Henry Falk

Abstract

This chapter provides insights into the health impacts and governance challenges related to climate change. It categorizes health effects into three main types: direct health effects from extreme weather conditions, indirect health effects due to changing environmental conditions, and complex pattern health effects influenced by various factors such as environmental, socioeconomic, and political elements. In addition, it draws attention to two critical cross-cutting issues: mental health impacts and the vulnerability of specific populations. The governance section explores responses to climate change at multiple levels, ranging from global to individual actions. Through global data, occurrences, and initiatives, with a particular focus on the USA, this chapter offers an exhaustive overview of the multifaceted health impacts of climate change and the corresponding governance strategies.

Keywords

Climate change · Indirect health effects · Governance · Extreme weather disasters · Mental health · Public health

1.1 Introduction

This chapter is an expanded version of two brief presentations made at the inaugural program of the Prof Rajani Prasanna Endowment Lecture Series on Climate Change and its Impact on Public Health held on 4 October 2021. This chapter adds to the lecture with more recent data, events, reports, and programs to provide an updated

H. Falk (✉)
Rollins School of Public Health, Emory University, Atlanta, GA, USA

H. Brand et al. (eds.), *Climate Change and Public Health*,
https://doi.org/10.1007/978-981-95-0895-2_1

and current view of the health effects and governance issues related to climate change. The Prof Rajani Prasanna Endowment Lecture Series is particularly timely—as I write this draft in June 2023, carbon dioxide levels in the air are at record highs, global temperatures for June 2023 are the highest on record for this time of year, the oceans are heating to record highs, and Antarctic Sea ice is falling to record lows for this time of year (Ramirez 2023b).

In this chapter, I start by discussing the different types of health effects of climate change and the associated cross-cutting issues of mental health and disproportional vulnerability. Next, I delineate the various governance arrangements in place to respond to the issue of climate change. Throughout, I have used global data, events, and programs for illustrative purposes, but I have in several instances particularly focused on the USA with which I am most familiar.

Health effects are categorized as 1. Direct health effects resulting from extreme weather conditions, both short and long term. 2. Indirect health effects when changing environmental conditions due to climate change led to effects on health. 3. Complex pattern health effects where broad climate, environmental, socioeconomic, political, and other factors combine to impact health. For this chapter, I have also included two other key cross-cutting issues under the complex pattern category. The first is mental health, which can be affected by virtually all the climate risks discussed in this chapter. The second is the issue of vulnerable populations; while the specific vulnerable groups will vary by risk factor, for all climate risks there will be vulnerable populations who suffer disproportionately higher risk. The governance discussion provides an overview of the structured response to climate change in descending order from the global response to those at the regional, national, state, city, community, and individual levels.

1.2 Part 1-Health

1.2.1 Direct Health Effects

1.2.1.1 Frequency and Severity of Extreme Weather Disasters

In the USA, the National Oceanic and Atmospheric Administration (NOAA) has been providing annual reports summarizing all major weather disasters since 1980. NOAA is also increasingly using its extensive data collection capabilities to support research and decision-making on weather and climate-related issues across the globe. Figure 1.1 shows one of the most widely disseminated maps in the USA, that of the annual billion-dollar weather and climate disasters, in this case for 2021 (Smith 2022). Figure 1.2 shows the cumulative cost (CPI-adjusted) of disasters for each year since 1980, and the costliest disaster years have been during the most recent two decades. The costs are primarily driven by the costliest individual disasters, which in the USA are the major hurricanes, with Hurricane Katrina in 2005 as the costliest hurricane and 2017 as the costliest year due to the combined effect of Hurricanes Harvey, Maria, and Irma. The NOAA data seems to suggest that weather-related disasters in the USA are becoming more frequent, costly, and damaging

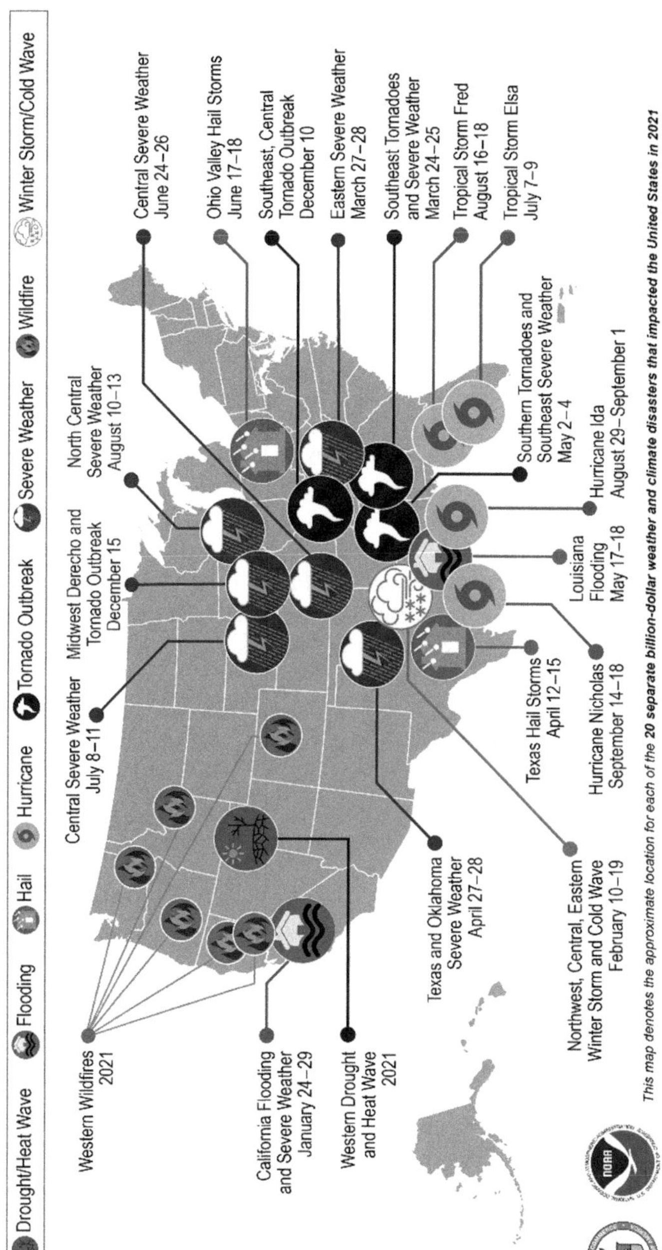

Fig. 1.1 US 2021-billion-dollar weather and climate disasters. (By Smith (2022). NOAA map by NOAA National Centre for Environmental Information (NCEI))

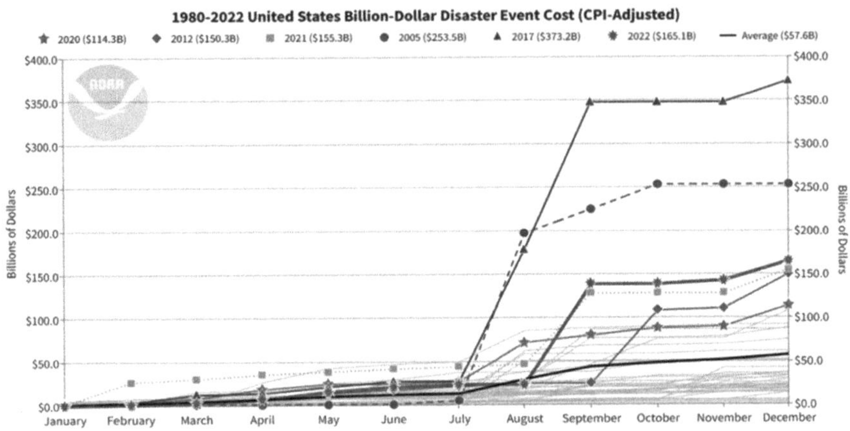

Fig. 1.2 Month-by-month accumulation of estimated costs of each year's US billion-dollar disasters, with colored lines showing 2022 (red) and the previous top-five costliest years. Other years are light gray. (By Smith (2022). NOAA map by NOAA National Centre for Environmental Information (NCEI) graphic)

(Smith 2022). Indeed, this is the public perception. A recent National Geographic article, in response to the disasters seen in Fig. 1.1, states this succinctly: "2021's weather disasters brought home the reality of climate change—Heat waves, floods, and megadroughts. This year's weather showed us that climate change is here and deadly" (Gibbens 2021).

However, the cost and economic impact of disasters depends on multiple factors, such as the severity of the disaster event, the density and vulnerability of the affected population, and the infrastructure development in the path of the disaster event (Jones et al. 2022). Cost alone is not a good measure of whether climate change is increasing weather-related disasters. Three approaches to address this issue are as follows:

1. IPCC modeling: The Intergovernmental Panel on Climate (IPCC 2021) has projected those incremental increases in global warming will lead to increased frequency and severity of major storms; Table 1.1 summarizes their projection for severe heat events. The IPCC effort is described in more detail in the governance section of this chapter.
2. World weather attribution: Using computer modeling and databases going back centuries or even millennia, the impact of current specific disasters can be compared to similar disasters in the past to ascertain whether the current disaster is more severe than would have been anticipated based on historical data (Voosen 2022). For example, Table 1.2 summarizes recent weather attribution presentations, listing six extreme weather events deemed all but impossible without global warming and four others that were made much more likely or more severe by climate change (Ramirez 2023a).
3. Epidemiological approach: This is perhaps not as objective as the above approaches, but by looking at the shifting pattern and nature of disaster events,

Table 1.1 IPCC projected changes in the frequency and intensity of extreme heat events (those occurring once every 10 or 50 years) at additional increments of global warming

10-year extreme heat events	Frequency per 10 years	Intensity increase
1850–1900	1×	0 °C
Present at +1 °C	2.8×	+1.2 °C
Future at +1.5 °C	4.1×	+1.9 °C
Future at +2 °C	5.6×	+2.6 °C
Future at +4 °C	9.4×	+5.1 °C
50-year extreme heat events	1×	0 °C
Present at +1 °C	4.8×	+1.2 °C
Future at +1.5 °C	8.6×	+2.0 °C
Future at +2 °C	13.9×	+2.7 °C
Future at +4 °C	39.2×	+5.3 °C

Adapted from IPCC (2021)

Table 1.2 Recent extreme weather events that would have been either extremely unlikely to have happened or would have been significantly less intense if not for climate change

A-Extremely unlikely to have happened:	
1. Siberian heat wave 2. Pacific Northwest heat wave	2020 2021
3. Northern Hemisphere drought	2022
4. Horn of Africa drought	2020–2023
5. Mediterranean drought	2023
6. Extreme heat in South Asia	2023
B-Likely to have been less severe: 7. Australia Black Summer fires	2019–2020
8. Western US drought 9. Hurricane Ian	2020–2023 2022
10. Pakistan floods	2022

Adapted from (Ramirez 2023b)

one can see clues suggesting that disaster events would have been extremely unlikely absent climate change. Examples such as the Pacific Northwest US/Canada 2021 heat wave and the huge flooding from Hurricane Harvey are discussed below.

1.2.1.2 Extreme Heat

In the USA, heat waves have traditionally occurred in older urban centers ("heat islands") in cities such as New York, Chicago, Philadelphia, and St Louis. As a result, the Pacific Northwest heat wave of 2021, in an area where air conditioning is not nearly as prevalent, was a wildly improbable event that shattered records (e.g., Lytton, Canada reached 121 °F, Portland OR 116 °F, and Seattle WA 108 °F) and left a clear impression that climate change had to have played a role (EPA 2022; Milman 2021). Of course, records were shattered elsewhere that summer, including in Sicily, Italy, which reached 119.8 °F, and in parts of the Middle East and Asia. In April 2023, a historic heat wave in Southeast Asia and China led to record temperatures, such as 114 °F in Thailand (Stillman 2023); Europe also saw the hottest April weather on record with a high of 101.8 °F in Spain (Livingston 2023a). Global projections predict even higher temperatures in the coming years (Plumer 2023). Electrical grid failure events ("blackouts") have been noted with increasing frequency during heat waves, at least partially due to high demand for air conditioning. A recent study estimated that a concurrent heat wave and blackout in Phoenix AZ, for example, would likely send nearly half of all Phoenix residents to the emergency room for treatment of heat stroke and related disorders (Stone et al. 2023). Extreme temperatures have been associated with elevations in multiple causes of disease, including metabolic diseases such as heat stroke, cardiovascular disease, suicide, and injuries (Ebi et al. 2021; The Lancet 2021). Several recent reviews discuss preparedness for extreme heat events (Hess et al. 2023; Jay et al. 2021).

Far more pervasive than sporadic heat waves, however, are the projections that prolonged extreme heat in many regions will have a major impact on health and socioeconomic conditions. A recent report suggested that by 2050 many megacities will resemble Kuwait City and Basra with 50–100 or more days a year with a heat index above 39 °C, with Kolkata, Karachi, Delhi, Manila, Lagos, and Guangzhou listed among the highest (Rubin et al. 2022). A World Bank report has projected the devastating economic impact of prolonged extreme heat conditions on living standards in central India (Mani et al. 2018), and a recent report suggests that global warming could see billions of people living in places with dangerously hot conditions where flourishing human life would be challenged (Lenton et al. 2023).

Wildfires in a sense are a direct extension of heat waves and extreme heat conditions. Recent decades have seen a surge in both frequency and severity of wildfires (Mansoor et al. 2022). The direct health effects are burns and other injuries, but the short- and long-term respiratory effects from widespread smoke damage among children and people with pre-existing chronic diseases are increasingly recognized (Finlay et al. 2012). The distribution of wildfires has been very broad (such as the US West, Australia, Alaska, Europe, Russia, and in May 2023 in Alberta, Canada). An extreme example in 2023 has been the record fires in Canada (Livingston 2023b) accompanied by dramatic headlines in major newspapers: *"How could this happen?": Canadian fires are burning where they rarely have before* (Onishi 2023); and *"The fire equivalent of an ice age": Humanity enters a new era of fire* (Osaka et al. 2023). New York City and other parts of the US Northeast and Midwest have suffered under major air pollution alerts as their skies were blanketed by wildfire smoke (Hirschlag 2023). With the dramatic increase in intensity, frequency, and magnitude of wildfires there are acutely heightened concerns about improved forest management, fire-fighting resources, the dangers to firefighters, and how to manage the urban-forest interface.

1.2.1.3 Storms, Floods, and Sea-Level Rise

Recent years have seen unprecedented flooding in multiple locations. In the summer of 2021, there were a series of extreme flood events in rapid succession; first in Germany/Belgium, followed by Zhengzhou, China, and then Tennessee and Hurricane Ida in the USA (Pierson et al. 2021). In 2022, the massive flood in Pakistan created a humanitarian disaster for more than six million people, with an overwhelming impact on food security and health systems (Devi 2022b); the Pakistan flood quickly became the most visible theme for the COP27 global climate meeting in Egypt a few months later. The year 2023 has already seen extraordinary floods in northern Italy (Povoledo 2023). A major hurricane with a very clear connection to climate change was the 2017 Hurricane Harvey in Texas where the storm meandered along the Texas coast near Houston for 5 days while constantly replenishing the rain clouds with warming waters from the Gulf of Mexico. Hurricane Katrina was an unprecedented flooding of a modern city, leading to a total evacuation and shutdown of all city functions (Pacheco 2017); restoring the city was a very complex operation to repair all the damaged infrastructure (Fig. 1.3).

The acute health effects of storms and floods include the following: lack of access to healthcare with the disruption to the healthcare system, particularly affecting the availability of medication and ongoing care for individuals with chronic diseases; injuries and deaths directly caused by the floods and the clean-up; lack of food and shelter; infectious diseases from crowding in shelters and lack of water/sanitation; and mold, which is ubiquitous in hot summers and can lead to respiratory illnesses (Du et al. 2010).

Rising seas due to climate change are a huge long-term threat to coastlines, major coastal cities, and low-lying inland areas. Compounding the effects of rising sea levels are demographic shifts to rapidly growing coastal megacities, the depletion of protective natural shorelines, land subsidence as aquifers are depleted, and due to the lack of barriers and adequate infrastructure for protection (Tay et al. 2022). The NOAA anticipates a sea-level rise of approximately one foot on average for US coastlines by 2050, leading to increased flooding, intrusion of saltwater, and other damaging effects (Fig. 1.4) (Sweet et al. 2022).

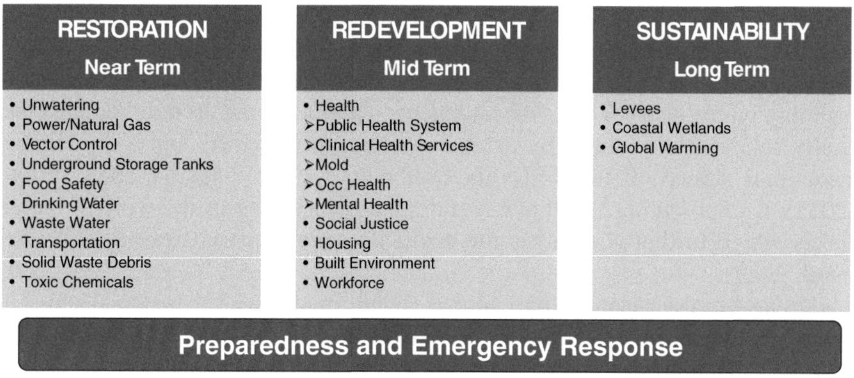

Fig. 1.3 Conceptual Framework for New Orleans. Centre for Disease Control and Prevention (CDC). (By Centre for Disease Control and Prevention (CDC), USA)

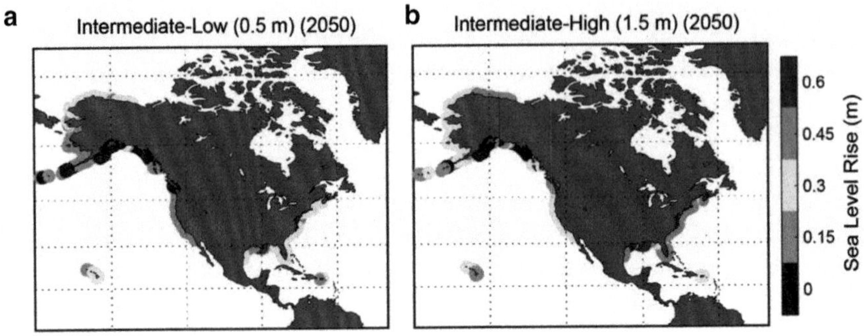

Fig. 1.4 Relative sea-level rise, in meters, in 2050 for the (**a**) Intermediate-Low, and (**b**) Intermediate-High scenarios relative to the year 2000. (By Sweet et al. (2022). National Oceanic and Atmospheric Administration (NOAA), USG)

The uncertain future of glaciers, particularly in Greenland and the Antarctic ice shelf, adds further risk (Rounce et al. 2023). Many coastal cities are already engaged in planning responses, many of which may be very costly (Yaro and Gutman 2023). Perhaps the most striking response is in Indonesia, which is responding to the severe flooding threat to its capital city Jakarta by planning to build a completely new capital in a more protected area (Paddock and Suhartono 2019).

1.2.1.4 Drought

At present, climate change is driving major parts of East Africa toward famine, and researchers for the World Weather Attribution initiative have concluded that this drought might not have happened without the impact of climate change (Devi 2022a; Thompson 2023); in general, climate-related water scarcity is a major challenge throughout Africa (Leal Filho et al. 2022). Many LMICs such as Syria and Iraq are similarly facing severe droughts, which are exacerbated by diminishing rainfall, decreased surface water, depleted aquifers, poor infrastructure, leakage from low-quality water storage and distribution systems, and the disruptive impact of conflict zones (Devi 2021). In recent years, several major cities such as Cape Town, South Africa, approached the catastrophe of running out of water. The US West has been amid an epic drought, at least partially relieved this winter by extensive atmospheric rivers and a replenished mountain snowpack in California (Samenow and Livingston 2023; Whyte 2022). Compounding the above is a dramatic shrinking of the world's largest lakes which further challenges the available water supply (Cooley 2023; Yao et al. 2023).

Drought is a substantial contributor to malnutrition, threats to maternal and child health, increases in infectious and chronic diseases, and complex pattern health effects from drought-induced population displacement, migration, poverty, and conflict zones (Salvador et al. 2023).

1.3 Indirect Health Effects

1.3.1 Vectors and Infectious Diseases

Climate change will lead to changes in habitats, geographic spread, and virulence of many vectors of infectious disease, with varied effects on pathogens, vectors, media, and hosts (Baker et al. 2021). In most cases, this will lead to increased risk in threatened populations, although in some instances the changes may be protective. The list of potentially affected infectious diseases is very long, including viral, parasitic, bacterial, and fungal diseases, and waterborne and foodborne diseases (Rao and Nagendra 2020). Perhaps the best studied to date is the impact on dengue and related viral diseases (Bhatia et al. 2022; Mondal 2023). A recent review has summarized the threats from malaria, dengue and zika viruses, Lyme Disease, and West Nile Virus infection (Thomson and Stanberry 2022).

1.3.1.1 Air Pollutants, Pollen, and Toxic Chemicals

Climate-induced changes in atmospheric patterns will lead to increases in air pollutants, particularly particulate matter. This will exacerbate existing challenges in countries already facing markedly elevated levels of air pollution, such as India. Air pollution primarily affects a variety of respiratory and cardiovascular diseases, but also affects reproductive outcomes and the central nervous system (Cohen et al. 2017; Rajagopalan and Landrigan 2021); it is a particular risk to vulnerable population groups such as children with chronic diseases (Perera and Nadeau 2022). Changing patterns of vegetation will affect pollen and respiratory irritants, leading to increases in asthma and allergic diseases. Environmental, industrial, agricultural, and population shifts will undoubtedly lead to new challenges in exposure to toxic chemicals, environmental hazards, pesticides, occupational hazards, and environmental disease.

1.3.1.2 Agriculture, Fisheries, and Livestock

Climate change will affect agricultural yields, the quality as well as quantity of food produced, and food distribution systems; these will necessarily impact the composition of diets and food safety, likely affecting nutritional quality and chronic diseases such as obesity and cardiovascular disease (Larkin 2022; Mahapatra et al. 2021). One striking example is the threat to rice growing, a dietary staple for billions of people (Sengupta et al. 2023). Fisheries and seafood production are also likely to diminish, further affecting diets. New agricultural approaches, governance arrangements, and technological innovations will be necessary to address these challenges (Dietz and Pryor 2022; Fanzo et al. 2022).

1.3.1.3 Complex Pattern Health Effects

Migration The United Nations estimates that there will be approximately one billion climate migrants by 2050; the UN Secretary-General recently warned that global warming could force a mass exodus “on a biblical scale” (Carrington 2023).

This will have a dramatic impact on the health and wellbeing of the displaced populations as well as on their host countries and highlights the need for innovative strategies to address this global threat (Balsari et al. 2020; McMichael 2023).

Conflict Dwindling access to resources such as food and water, large-scale migration, and the resulting health and socioeconomic impacts will inevitably lead to conflicts such as social upheaval, civil disturbances, and war (Hendrix et al. 2023). These will have enormous consequences on global health as we are currently seeing in countries such as Ukraine and Syria.

Poverty This is a likely result of all the direct, indirect, and complex pattern effects detailed above. Poverty itself results in large-scale health inequities (Callahan and Mankin 2023).

Mental Health Virtually all major disasters and societal upheavals lead to mental health impacts such as PTSD, particularly affecting vulnerable populations such as children and those with underlying mental health illnesses (Falk 2013).

Vulnerable Populations Different groups are disproportionately affected by the various direct, indirect, and complex pattern health effects described above (EPA 2021). In general, vulnerable populations include the following:

- Individuals with disabilities and/or chronic diseases—socioeconomically deprived or impoverished groups.
- Marginalized communities, such as those resulting from environmental degradation or racism.
- Vulnerable age groups, such as children, pregnant women, and the elderly.
- Occupational groups exposed to major heat stress.
- Indigenous populations, such as Native Americans in the USA.
- Geographically exposed populations, such as coastal populations affected by rising seas.
- The global slum population is estimated to grow to two billion people by 2030 (Ross et al. 2019).

However, the specific patterns of vulnerability associated with each of the health impacts of climate change are likely to be different and context dependent.

1.4 Part 2-Governance

1.4.1 Global

Perhaps the single most critical function at the global level is tracking the climate changes resulting from global warming. This function is carried out by the Intergovernmental Panel on Climate Change (IPCC) which was established as a joint venture of the World Meteorological Organization (WMO) and the United Nations Environment Program (UNEP). The first IPCC report was published in 1990; the sixth cycle of assessment reports (AR6) started in 2021 (IPCC 2021) and was concluded in mid-2023 (IPCC 2023). The best known of these reports is the Physical Science Report which focuses on climate conditions and projections (IPCC 2021). The reports have become increasingly definitive over the past three decades, and their conclusions underpin all the mitigation and adaptation responses to global warming. Several key conclusions from the AR6 report in 2021 follow:

1. "...unequivocal that human influence has warmed the atmosphere, ocean, and land."
2. "Observed increases...are unequivocally caused by human activities."
3. "Each of the last four decades has been successively warmer than any decade that preceded it."
4. "Human influence is very likely the main driver of the global retreat of glaciers since the 1990s."

The AR6 Synthesis Report released in March 2023 stated that the world is likely to surpass the 1.5 °C climate target by the early 2030s leaving it on the brink of catastrophic warming (IPCC 2023).

The WMO and UNEP also produce their annual reports on global climate conditions. The WMO provides detailed climate data for the year, including concentrations of major GHGs, global mean temperature, global sea-level rise, ocean heat content, and information about global weather disasters for that year (WMO 2022). The UNEP Adaptation Gap Report assesses the shortfall between announced country pledges and plans for emission reductions and what is happening (UNEP 2022). Their most recent and very blunt assessment: "Too little, too slow: Climate adaptation future puts the world at risk," a conclusion shared by many others (Matthews and Wynes 2022; Rogelj et al. 2023).

A critical part of the global response to climate change is the UN Framework Convention on Climate Change (UNFCCC), which began in 1994 and has been ratified by 197 countries (UNFCCC n.d.). This is the framework through which countries make national pledges at the annual meetings to reduce emissions and to provide support for global emission reduction efforts. Key outcomes from the UN Climate Change Conference—COP26—in Glasgow, UK, in 2021 included major efforts on funding to assist developing countries and on phasing out the use of coal; the WHO and others advanced the theme of framing climate change as a health issue (Balbus et al. 2022). COP27 in Sharm el-Sheikh, Egypt, in 2022 was unfortunately impacted by the Russia-Ukraine war and its many effects on global energy, food, and the economy. Nevertheless, it approved a fund for LMIC spurred by the overwhelming flooding in Pakistan. There was much discussion on extending the phaseout of coal to oil and gas but without agreement. A Lancet report (Siva 2022) described COP27 as a collective failure, an opinion held by many; *The Economist* cover story was titled "Say Goodbye to 1.5 °C—Why climate policy is off target"

(The Economist 2022). COP28 in 2023 will be held in the UAE; there are already concerns that it will be unduly influenced by the petrochemical industry.

Several of the leading global philanthropies and foundations, such as Gates, Bloomberg, and the Wellcome Trust are strong advocates as well as leaders in promoting effective action on climate change. In the private sector, the automotive industry appears to be shifting aggressively to the production of electric vehicles, with government regulations as in the USA often promoting the change. Many global scientific, academic, and professional groups are also actively pursuing climate change goals (e.g., *The New England Journal of Medicine*) (NEJM 2021). A leader in this area, for example, is the British medical journal *Lancet* which established a Countdown on Health and Climate Change; it utilized a very large group of leading researchers to create a detailed set of indicators (44 in the 2021 version) to both advocate for and track progress in climate change and health (Romanello et al. 2022; Watts et al. 2018).

1.4.1.1 Regional

There are multiple regional efforts underway. For example, Southeast Asia is facing very significant threats such as the erosion of the Mekong River Delta and looming challenges to rice production, a staple of the Asian diet (Tamura et al. 2020). Both the Asian Development Bank and the Asia-Pacific Regional Forum on Health and Development support climate-related initiatives (Gilfillan 2018). At COP27, President Herzog of Israel proposed the creation of "The Renewable Middle East"; he stated that the Middle East not just be seen as a vast desert that is warming faster than other parts of the planet, but as a "ray of sustainably powered light" with a vast potential for clean energy (Linares et al. 2020; Neira et al. 2023). At the same time, there are many looming regional challenges. For example, as the Arctic region warms and as the Arctic Ocean becomes more accessible and navigable, issues regarding boundaries, access to resources, control, ownership, pollution, etc., are formidable and have the potential to lead to conflict (Fischetti 2019).

1.4.1.2 National

There are many difficult challenges to developing a consistent and coherent national policy on an issue as complex as climate change. First, many urgent realities compete for government responses such as the COVID-19 pandemic and the Russia-Ukraine war that at a minimum distract from climate change policy development. More recently, widening political divisions in many countries have made collective action more difficult. Even when firm policies are established, they often do not survive the next election cycle and can be reversed by a new party in power. Also, there is a governance challenge of creating the optimum administrative structure for developing a government-wide response to a challenge as pervasive as climate change (Dubash et al. 2021). Examples of these challenges abound.

In the USA, there was a complete turnaround in climate policy from Obama to Trump to Biden (Martins 2021; Wallach 2019); this will likely continue for some time. In Germany, the government was lauded for moving away from coal and nuclear power at the beginning of 2022; the circumstances of European energy policy after the Russian invasion of Ukraine created a whole new reality for energy policy (Connolly 2022). In Guyana, the government is in the unusual circumstance of developing huge new offshore oil fields to pay for the extensive sea wall necessary to protect their below-sea-level population centers affected by sea-level rise caused substantially by climate change (Juhasz 2021). Madagascar, an impoverished African nation that has been devastated by multiple cyclones, drought, and deforestation, has minimal resources with which to even develop a sound response plan (USAID n.d.). In Pakistan, the severe flood of 2022 is an acute disaster that only exacerbates the ongoing and severe long-standing challenges related to safe and adequate water resources for the country (Ahmed et al. 2020).

Pakistan is among the countries most affected by climate change, and it has a rapidly growing population with increasing demand for water resources; there are major concerns about water quantity, quality, and waterborne diseases; there are major deficiencies in the protection of water resources; and climate change will greatly affect its glaciers, the monsoons, and the Indus River water supply. There is a clear need for national water planning which is far more complex than just the recent floods, as damaging and pervasive as they were (Ahmed et al. 2020; Bhutta et al. 2022). Similar concerns are also likely to be true for parts of India, given that it shares many of the water sources as well as challenges with Pakistan.

In the USA, the Biden administration has come forward with many new policies and funding decisions related to climate change:

1. Created a new senior-level, Cabinet position for climate leadership.
2. Created a new Office for Climate Change, Health, and Equity in the Department of Health and Human Services.
3. Passed a bipartisan USD 1 trillion infrastructure bill with USD 50 billion in climate funds.
4. Passed the Inflation Reduction Act with USD 360+ billion for climate funds (see below).
5. Promulgated multiple Agency regulations to support electric vehicles and renewable energy sources.
6. Set a US target of reaching carbon neutrality by 2050.
7. Developed climate plans for each of the 21 US Agencies.
8. Provided support for global initiatives, for example, Biden participated in both COP26 and 27.

The Inflation Reduction Act is perhaps the most substantial of the above given the extraordinary amount of funding (Glicksman 2023; Jaffe 2022). The unusual name means that the legislation had to pay for itself by raising resources to cover climate funding. Of note, legislation is much more difficult to undo than Executive Orders, so this has staying power for the long term. Also, many of the provisions will last for up to 10 years, so industry and others have time to plan for the use of the funds. The hope is that this legislation will lead to a 31–44 percent reduction in US GHG emissions by 2030 instead of the currently anticipated reduction of 24–35 percent. The legislation includes the following:

1. Clean energy tax credits.
2. Clean electricity investment tax credits.
3. Fuel tax credits, clean vehicle tax credits, and credits for a plethora of actions such as carbon management, low-carbon construction materials, and building improvements.
4. Funds for energy innovation in high-emission industries.
5. Funding for the national labs, offshore wind and energy, and methane emission reduction.
6. Community investment with funds for energy justice and low-income and rural communities.
7. Funds for agriculture, forestry management, ports, etc.
8. A Department of Energy loan program and Green Banks to support climate initiatives.

1.4.1.3 States/Provinces and Cities

In the USA, India, and many other countries, national policies provide direction and guidance, but many decisions regarding funding and implementation of programs and projects are made at the state (provincial) and local levels. I alluded to this earlier when discussing costly programs being designed at the state and local level to protect major US coastal cities from flooding and rising sea levels. There are complex governance relationships between state and city governments that vary widely by locale and are beyond the scope of this chapter. However, cities worldwide have become a key focus of innovation at the nexus of climate change and urban planning, and I provide a brief introduction to this topic here.

Urbanization and the growth of cities have surged over the past century. Cities have grown from 10 percent of the world's population at the beginning of the twentieth century, to 29 percent in 1950, and more than 50 percent in 2022, and are projected to be close to 70 percent by 2050 (Adlakha and John 2022; Wigginton et al. 2016). This means that about 2.5 billion more people will be living in cities by 2050 (UN et al. 2019). The central importance of cities to climate change is clear: They produce up to 70 percent of global carbon dioxide emissions, use vast quantities of water, and produce mountains of waste (McDonough 2017). Because of their central role, progress in cities will also be crucial in achieving success for the UN Sustainable Development Goals (SDGs) (Giles-Corti et al. 2020).

It is also clear that siloed efficiencies or improvements in individual sectors (e.g., reductions in air pollution or solid waste) are beneficial but not adequate for major change. Cities in the past were largely designed for linear flow: "Stuff" such as metal, plastic, and food come in at one end, are used, and then wastes are discarded to landfills, incinerators, etc., at the other end; among many other things, this is extremely wasteful of resources and energy. Urban planners and climate change experts talk of reimagining cities, where wastes become resources for future use, renewable energy is maximized, and there is a much greater focus on benefiting to the extent possible from local ecology, geology, circumstances, etc. To do this well requires extensive interaction among many stakeholders (such as health, energy,

Table 1.3 Listing of key individual/household and community-level actions to respond to climate change

Individual/household/community initiative	
Individual/household	Community
Spread the word Maintain political pressure ↓ AC and power Use Shop local and buy sustainably Leave the car at home Invest responsibly Cut consumption and waste Dress smart Plant trees Cut back on flying	Protect green spaces Transform transport Advocate to those in power Smart land-use & zoning decisions Better building strategies Promote climate-friendly policies Action plans for nursing homes, daycare facilities, and schools Promote climate-friendly hospitals and other local institutions

transportation, commerce, water, and food) (Cities Alliance and UNOPS 2022; UN-Habitat 2022).

Two major networking alliances exist for cities focusing on climate change: the Cities Alliance (https://citiesalliance.org) and the C40 global network of mayors (https://www.c40.org). The websites describe an extraordinary array of growing networks, training programs, and joint accomplishments (Alliance 2023; C40 2023).

1.4.1.4 Communities, Households, and Individuals

While this chapter has focused on governance and the institutional structures for governance, it is readily evident to anyone familiar with climate change that so much vision, energy, initiative, and progress related to climate change have risen from the ground up. Table 1.3 provides some very clear examples of initiatives that can be taken by all individuals and households and by their local communities. This is an issue where self-reliance and community actions have been paramount in driving action and resources. The scale of impacts as outlined in this chapter makes it clear that each of us will have to do our part in rising to the unprecedented challenge of climate change.

References

Adlakha D, John F. The future is urban: integrated planning policies can enable healthy and sustainable cities. Lancet Glob Health. 2022;10(6):e790–1. https://doi.org/10.1016/S2214-109X(22)00211-X.

Ahmed T, Zounemat-Kermani M, Scholz M. Climate change, water quality and water-related challenges: a review with focus on Pakistan. Int J Environ Res Public Health. 2020;17(22):1–22. https://doi.org/10.3390/IJERPH17228518.

Alliance C. Homepage | Cities Alliance. Cities Alliance, 2023. https://www.citiesalliance.org/.

Baker RE, Mahmud AS, Miller IF, Rajeev M, Rasambainarivo F, Rice BL, Takahashi S, Tatem AJ, Wagner CE, Wang LF, Wesolowski A, Metcalf CJE. Infectious disease in an era of global change. Nat Rev Microbiol. 2021;20(4):193–205. https://doi.org/10.1038/s41579-021-00639-z.

Balbus JM, McCannon CJ, Mataka A, Levine RL. After COP26 — putting health and equity at the Center of the Climate Movement. N Engl J Med. 2022;386(14):1295–7. https://doi.org/10.1056/NEJMP2118259/SUPPL_FILE/NEJMP2118259_DISCLOSURES.PDF.

Balsari S, Dresser C, Leaning J. Climate change, migration, and civil strife. Curr Environ Health Rep. 2020;7(4):404–14. https://doi.org/10.1007/S40572-020-00291-4/TABLES/3.

Bhatia S, Bansal D, Patil S, Pandya S, Ilyas QM, Imran S. A retrospective study of climate change affecting dengue: evidences, challenges and future directions. Front Public Health. 2022;10:884645. https://doi.org/10.3389/FPUBH.2022.884645/BIBTEX.

Bhutta ZA, Bhutta SZ, Raza S, Sheikh AT. Addressing the human costs and consequences of the Pakistan flood disaster. Lancet. 2022;400(10360):1287–9. https://doi.org/10.1016/S0140-6736(22)01874-8.

C40. C40 Cities - A global network of mayors taking urgent climate action. C40 Cities Climate Leadership Group, 2023. https://www.c40.org/.

Callahan CW, Mankin JS. Persistent effect of El Niño on global economic growth. Science. 2023;380(6649):1064–9. https://doi.org/10.1126/SCIENCE.ADF2983.

Carrington D. Rising seas threaten 'mass exodus on a biblical scale', UN chief warns. The Guardian, 2023, February 14. https://www.theguardian.com/environment/2023/feb/14/rising-seas-threaten-mass-exodus-on-a-biblical-scale-un-chief-warns.

Cities Alliance, & UNOPS. Seizing the opportunity: rapid urbanisation and the circular economy at the intersection of climate change and poverty. In Cities Alliance, 2022.

Cohen AJ, Brauer M, Burnett R, Anderson HR, Frostad J, Estep K, Balakrishnan K, Brunekreef B, Dandona L, Dandona R, Feigin V, Freedman G, Hubbell B, Jobling A, Kan H, Knibbs L, Liu Y, Martin R, Morawska L, et al. Estimates and 25-year trends of the global burden of disease attributable to ambient air pollution: an analysis of data from the Global Burden of Diseases Study 2015. Lancet. 2017;389(10082):1907–18. https://doi.org/10.1016/S0140-6736(17)30505-6.

Connolly K. Germany to reactivate coal power plants as Russia curbs gas flow | Germany | The Guardian. The Guardian, 2022, July 8. https://www.theguardian.com/world/2022/jul/08/germany-reactivate-coal-power-plants-russia-curbs-gas-flow.

Cooley SW. Global loss of lake water storage. Science (New York, NY). 2023;380(6646):693. https://doi.org/10.1126/SCIENCE.ADI0992.

Devi S. Syria and Iraq facing severe drought. Lancet (London, England). 2021;398(10309):1395. https://doi.org/10.1016/S0140-6736(21)02238-8.

Devi S. Climate change driving East Africa towards famine. Lancet (London, England). 2022a;400(10347):150–1. https://doi.org/10.1016/S0140-6736(22)01325-3.

Devi S. Pakistan floods: impact on food security and health systems. Lancet (London, England). 2022b;400(10355):799–800. https://doi.org/10.1016/S0140-6736(22)01732-9.

Dietz WH, Pryor S. How can we act to mitigate the global Syndemic of obesity, undernutrition, and climate change? Curr Obes Rep. 2022;11(3):61–9. https://doi.org/10.1007/S13679-021-00464-8.

Du W, Fitzgerald GJ, Clark M, Hou XY. Health impacts of floods. Prehosp Disaster Med. 2010;25(3):265–72. https://doi.org/10.1017/S1049023X00008141.

Dubash NK, Pillai AV, Flachsland C, Harrison K, Hochstetler K, Lockwood M, MacNeil R, Mildenberger M, Paterson M, Teng F, Tyler E. National climate institutions complement targets and policies; institutions can affect coordination, consensus, and strategy. Science. 2021;374(6568):690–3. https://doi.org/10.1126/SCIENCE.ABM1157/SUPPL_FILE/SCIENCE.ABM1157_SM.PDF.

Ebi KL, Capon A, Berry P, Broderick C, de Dear R, Havenith G, Honda Y, Kovats RS, Ma W, Malik A, Morris NB, Nybo L, Seneviratne SI, Vanos J, Jay O. Hot weather and heat extremes: health risks. Lancet. 2021;398(10301):698–708. https://doi.org/10.1016/S0140-6736(21)01208-3.

EPA. Climate change and social vulnerability in the United States: a focus on six impacts, 2021.

EPA. Climate change indicators: heat waves | US EPA. United States Environmental Protection Agency, 2022, August 1. https://www.epa.gov/climate-indicators/climate-change-indicators-heat-waves.

Falk H. Natural disasters, environmental emergencies, and children's health. In: Landrigan PJ, Etzel RA, editors. Textbook of children's environmental health. Oxford University Press; 2013. p. 546–58. https://doi.org/10.1093/MED/9780199929573.003.0059.

Fanzo J, Rudie C, Sigman I, Grinspoon S, Benton TG, Brown ME, Covic N, Fitch K, Golden CD, Grace D, Hivert MF, Huybers P, Jaacks LM, Masters WA, Nisbett N, Richardson RA, Singleton CR, Webb P, Willett WC. Sustainable food systems and nutrition in the 21st century: a report from the 22nd annual Harvard nutrition obesity symposium. Am J Clin Nutr. 2022;115(1):18–33. https://doi.org/10.1093/AJCN/NQAB315.

Finlay SE, Moffat A, Gazzard R, Baker D, Murray V. Health impacts of wildfires. PLoS Curr. 2012;4 https://doi.org/10.1371/4F959951CCE2C.

Fischetti M. Special report: future of the Arctic - Scientific American, 2019.

Gibbens S. 2021's weather disasters brought home the reality of climate change. National Geographic, 2021, December 7. https://www.nationalgeographic.com/environment/article/this-year-extreme-weather-brought-home-reality-of-climate-change.

Giles-Corti B, Lowe M, Arundel J. Achieving the SDGs: evaluating indicators to be used to benchmark and monitor progress towards creating healthy and sustainable cities. Health Policy (Amsterdam, Netherlands). 2020;124(6):581–90. https://doi.org/10.1016/J.HEALTHPOL.2019.03.001.

Gilfillan D. Regional organisations supporting health sector responses to climate change in Southeast Asia. Glob Health. 2018;14(1):1–13. https://doi.org/10.1186/S12992-018-0388-Z/TABLES/2.

Glicksman RL. Protecting the public health with the inflation reduction act - provisions affecting climate change and its health effects. N Engl J Med. 2023;388(1):84–8. https://doi.org/10.1056/NEJMMS2212981.

Hendrix CS, Koubi V, Selby J, Siddiqi A, von Uexkull N. Climate change and conflict. Nat Rev Earth Environ. 2023;4(3):144–8. https://doi.org/10.1038/s43017-022-00382-w.

Hess JJ, Errett NA, McGregor G, Isaksen TB, Wettstein ZS, Wheat SK, Ebi KL. Public health preparedness for extreme heat events. Annu Rev Public Health. 2023;44(1):301–21. https://doi.org/10.1146/ANNUREV-PUBLHEALTH-071421-025508.

Hirschlag A. The long distance harm to health caused by wildfires - BBC Future. BBC, 2023, June 7. https://www.bbc.com/future/article/20200821-how-wildfire-pollution-may-be-harming-your-health.

IPCC. Summary for policymakers. In: Masson-Delmotte V, Zhai P, Pirani A, Connors SL, Péan C, Berger S, Caud N, Chen Y, Goldfarb L, Gomis MI, Huang M, Leitzell K, Lonnoy E, Matthews JBR, Maycock TK, Waterfield T, Yelekçi O, Yu R, Zhou B, editors. Climate change 2021: the physical science basis. Contribution of Working Group I to the sixth assessment report of the intergovernmental panel on climate change. Cambridge University Press; 2021. p. 3–32. https://doi.org/10.1017/9781009157896.001.

IPCC. Summary for policymakers. In: The Core Writing Team, Lee H, Romero J, editors. AR6 synthesis report: climate change 2023. Contribution of Working Groups I, II and III to the sixth assessment report of the Intergovernmental Panel on Climate Change. Intergovernmental Panel on Climate Change; 2023. p. 1–34. https://doi.org/10.59327/IPCC/AR6-9789291691647.001.

Jaffe S. Health organisations welcome US climate crisis law. Lancet (London, England). 2022;400(10354):719–20. https://doi.org/10.1016/S0140-6736(22)01663-4.

Jay O, Capon A, Berry P, Broderick C, De Dear R, Havenith G, Honda Y, Kovats S, Ma W, Malik A, Morris NB, Nybo L, Seneviratne SI, Vanos J, Ebi KL. Heat and Health 2 reducing the health effects of hot weather and heat extremes: from personal cooling strategies to green cities. Lancet. 2021;398:709–24. https://doi.org/10.1016/S0140-6736(21)01209-5.

Jones RL, Guha-Sapir D, Tubeuf S. Human and economic impacts of natural disasters: can we trust the global data? Sci Data. 2022;9(1):1–7. https://doi.org/10.1038/s41597-022-01667-x.

Juhasz A. Exxon's oil drilling gamble off Guyana coast 'poses major environmental risk' | Environment | The Guardian. The Guardian, 2021, August 17. https://www.theguardian.com/environment/2021/aug/17/exxon-oil-drilling-guyana-disaster-risk.

Larkin H. Conflict and climate collide to create an acute hunger crisis for an unprecedented 345 million people. 2022, JAMA;328(22):2198–200. https://doi.org/10.1001/JAMA.2022.19235.

Leal Filho W, Totin E, Franke JA, Andrew SM, Abubakar IR, Azadi H, Nunn PD, Ouweneel B, Williams PA, Simpson NP. Understanding responses to climate-related water scarcity in Africa. Sci Total Environ. 2022;806(1):1–18. https://doi.org/10.1016/J.SCITOTENV.2021.150420.

Lenton TM, Xu C, Abrams JF, Ghadiali A, Loriani S, Sakschewski B, Zimm C, Ebi KL, Dunn RR, Svenning JC, Scheffer M. Quantifying the human cost of global warming. Nat Sustain. 2023;2023:1–11. https://doi.org/10.1038/s41893-023-01132-6.

Linares C, Díaz J, Negev M, Martínez GS, Debono R, Paz S. Impacts of climate change on the public health of the Mediterranean Basin population - current situation, projections, preparedness and adaptation. Environ Res. 2020;182:1–14. https://doi.org/10.1016/J.ENVRES.2019.109107.

Livingston I. Summerlike heat is baking Southern Europe as April records fall. The Washington Post, 2023a, April 27. https://www.washingtonpost.com/weather/2023/04/27/spain-portugal-record-april-heat/.

Livingston I. Why Canada's wildfires are extreme and getting worse, in 4 charts. The Washington Post, 2023b, June 12. https://www.washingtonpost.com/weather/2023/06/12/canada-record-wildfire-season-statistics/.

Mahapatra B, Walia M, Rao CAR, Raju BMK, Saggurti N. Vulnerability of agriculture to climate change increases the risk of child malnutrition: evidence from a large-scale observational study in India. PLoS One. 2021;16(6):1–15. https://doi.org/10.1371/JOURNAL.PONE.0253637.

Mani M, Bandyopadhyay S, Chonabayashi S, Markandya A, Mosier T. South Asia's hotspots: the impact of temperature and precipitation changes on living standards. World Bank; 2018. https://doi.org/10.1596/978-1-4648-1155-5.

Mansoor S, Farooq I, Kachroo MM, Mahmoud AED, Fawzy M, Popescu SM, Alyemeni MN, Sonne C, Rinklebe J, Ahmad P. Elevation in wildfire frequencies with respect to the climate change. J Environ Manag. 2022;301:1–7. https://doi.org/10.1016/J.JENVMAN.2021.113769.

Martins A. India-EU cooperation on climate change: converging or diverging interests? In: Inamdar N, Poojary PV, Shetty P, editors. Contours of India - EU engagements: multiplicity of experiences. Manipal Universal Press; 2021. p. 1–480.

Matthews HD, Wynes S. Current global efforts are insufficient to limit warming to 1.5°C. Science (New York, NY). 2022;376(6600):1404–9. https://doi.org/10.1126/SCIENCE.ABO3378.

McDonough W. How cities could save us: urban areas can improve the planet as well as people's lives if we design them to be much more resourceful with energy, water, food and minerals. Sci Am. 2017;317(1):44–8. https://doi.org/10.1038/SCIENTIFICAMERICAN0717-44.

McMichael C. Climatic and environmental change, migration, and health. Annu Rev Public Health. 2023;44(1):171–91. https://doi.org/10.1146/ANNUREV-PUBLHEALTH-071421-045148.

Milman O. Urban heat island effect exacerbating summer heatwaves, study shows | Climate crisis. The Guardian, 2021, July 14. https://www.theguardian.com/environment/2021/jul/14/urban-heat-island-effect-heatwave.

Mondal N. The resurgence of dengue epidemic and climate change in India. Lancet. 2023;401(10378):727–8. https://doi.org/10.1016/S0140-6736(23)00226-X.

Neira M, Erguler K, Ahmady-Birgani H, Daif Allah AL-Hmoud N, Fears R, Gogos C, Hobbhahn N, Koliou M, Kostrikis LG, Lelieveld J, Majeed A, Paz S, Rudich Y, Saad-Hussein A, Shaheen M, Tobias A, Christophides G. Climate change and human health in the Eastern Mediterranean and Middle East: literature review, research priorities and policy suggestions. Environ Res. 2023;216:1–23. https://doi.org/10.1016/J.ENVRES.2022.114537.

NEJM. Climate change articles: the New England Journal of Medicine. N Engl J Med. 2021; https://www.nejm.org/climate-change

Onishi N. Canadian wildfires are burning where they rarely have before - The New York Times. The New York Times, 2023, June 8. https://www.nytimes.com/2023/06/09/world/canada/quebec-canada-wildfires-locations.html.

Osaka S, Miller ME, Rios B. Canadian wildfires have ushered in a new fire era, the 'Pyrocene' - The Washington Post. The Washington Post, 2023, June 10. https://www.washingtonpost.com/climate-environment/2023/06/10/canadian-wildfire-smoke-pyrocene/.

Pacheco SE. Hurricane Harvey and climate change: the need for policy to protect children. Pediatr Res. 2017;83(1):9–10. https://doi.org/10.1038/pr.2017.280.

Paddock RC, Suhartono M. Indonesia's capital is sinking, polluted and crowded. Its leader wants to build a new one. The New York Times, 2019, August 26. https://www.nytimes.com/2019/08/26/world/asia/indonesia-capital-jakarta-borneo.html.

Perera F, Nadeau K. Climate change, fossil-fuel pollution, and children's health. N Engl J Med. 2022;386(24):2303–14. https://doi.org/10.1056/NEJMRA2117706.

Pierson D, Su A, Hennessy-Fiske M. Climate change ramps up heat, fires, and floods worldwide. Los Angeles Times, 2021, July 21. https://www.latimes.com/world-nation/story/2021-07-21/extreme-weather-worldwide-climate-change-disasters.

Plumer B. Heat will likely soar to record levels in next 5 years, WMO Says. The New York Times, 2023, May 17. https://www.nytimes.com/2023/05/17/climate/record-heat-forecast.html.

Povoledo E. 'Catastrophic' floods in Italy leave 8 dead and thousands homeless - The New York Times. The New York Times, 2023, May 17. https://www.nytimes.com/2023/05/17/world/europe/italy-flooding-emilia-romagna.html

Rajagopalan S, Landrigan PJ. Pollution and the heart. N Engl J Med. 2021;385(20):1881–92. https://doi.org/10.1056/NEJMRA2030281/SUPPL_FILE/NEJMRA2030281_DISCLOSURES.PDF.

Ramirez R. Without climate change, these extreme weather events would not have happened | CNN. CNN, 2023a, June 16. https://edition.cnn.com/2023/06/15/world/extreme-weather-events-climate-change/index.html.

Ramirez R. Four alarming charts that show just how extreme the climate is right now. CNN, 2023b, June 17. https://edition.cnn.com/2023/06/17/world/four-climate-charts-extreme-weather-heat-oceans/index.html.

Rao NV, Nagendra H. Epidemics and climate change in India. Curr Sci. 2020;119(12):1919–26.

Rogelj J, Fransen T, Elzen MGJ, den, Lamboll RD, Schumer C, Kuramochi T, Hans F, Mooldijk S, Portugal-Pereira J. Credibility gap in net-zero climate targets leaves world at high risk. Science. 2023;380(6649):1014–6. https://doi.org/10.1126/SCIENCE.ADG6248.

Romanello M, Di Napoli C, Drummond P, Green C, Kennard H, Lampard P, Scamman D, Arnell N, Ayeb-Karlsson S, Ford LB, Belesova K, Bowen K, Cai W, Callaghan M, Campbell-Lendrum D, Chambers J, van Daalen KR, Dalin C, Dasandi N, et al. The 2022 report of the Lancet Countdown on health and climate change: health at the mercy of fossil fuels. Lancet. 2022;400(10363):1619–54. https://doi.org/10.1016/S0140-6736(22)01540-9/ATTACHMENT/D63703F8-315E-4CDB-9573-1E552E1D4913/MMC5.PDF.

Ross AGP, Zaman K, Clemens JD. Health concerns in urban slums: a glimpse of things to come? JAMA. 2019;321(20):1973–4. https://doi.org/10.1001/JAMA.2019.3774.

Rounce DR, Hock R, Maussion F, Hugonnet R, Kochtitzky W, Huss M, Berthier E, Brinkerhoff D, Compagno L, Copland L, Farinotti D, Menounos B, McNabb RW. Global glacier change in the 21st century: every increase in temperature matters. Science. 2023;379(6627):78–83. https://doi.org/10.1126/SCIENCE.ABO1324/SUPPL_FILE/SCIENCE.ABO1324_SM.PDF.

Rubin AJ, Hubbard B, Holder J, Throop N, Rhyne E, White J, Glanz J. Extreme heat will change us. The New York Times, 2022, November 18. https://www.nytimes.com/interactive/2022/11/18/world/middleeast/extreme-heat.html.

Salvador C, Nieto R, Vicente-Serrano SM, García-Herrera R, Gimeno L, Vicedo-Cabrera AM. Public health implications of drought in a climate change context: a critical review. Annu Rev Public Health. 2023;44(1):213–32. https://doi.org/10.1146/ANNUREV-PUBLHEALTH-071421-051636.

Samenow J, Livingston I. California's snowpack soars to record high after 17 atmospheric rivers - The Washington Post, 2023, March 30. https://www.washingtonpost.com/weather/2023/03/30/california-snowpack-record-sierra/.

Sengupta S, Thuy TL, Nguyen T, Doyle R. How global rice farming Is being transformed by climate change. The New York Times, 2023, May 20. https://www.nytimes.com/interactive/2023/05/20/climate/rice-farming-climate-change.html.

Siva N. COP27: a "collective failure.". Lancet (London, England). 2022;400(10366):1835. https://doi.org/10.1016/S0140-6736(22)02407-2.

Smith A. 2021 U.S. billion-dollar weather and climate disasters in historical context. NOAA Climate.Gov, 2022, January 24. https://www.climate.gov/news-features/blogs/beyond-data/2021-us-billion-dollar-weather-and-climate-disasters-historical.

Stillman D. Historic Asia heat breaks hundreds of records; extremes in Thailand and China The Washington Post, 2023, April 17. https://www.washingtonpost.com/weather/2023/04/17/historic-heat-wave-asia-thailand/.

Stone B, Gronlund CJ, Mallen E, Hondula D, O'Neill MS, Rajput M, Grijalva S, Lanza K, Harlan S, Larsen L, Augenbroe G, Krayenhoff ES, Broadbent A, Georgescu M. How blackouts during heat waves amplify mortality and morbidity risk. Environ Sci Technol. 2023;57:8245–55. https://doi.org/10.1021/ACS.EST.2C09588/ASSET/IMAGES/LARGE/ES2C09588_0005.JPEG.

Sweet WV, Hamlington BD, Kopp RE, Weaver CP, Barnard PL, Bekaert D, Brooks W, Craghan M, Dusek G, Frederikse T, Garner G, Genz AS, Krasting JP, Larour E, Marcy D, Marra JJ, Obeysekera J, Osler M, Pendleton M, et al. Global and regional sea level rise scenarios for the United States. NOAA Technical Report NOS; 2022.

Tamura T, Nguyen VL, Ta TKO, Bateman MD, Gugliotta M, Anthony EJ, Nakashima R, Saito Y. Long-term sediment decline causes ongoing shrinkage of the Mekong megadelta, Vietnam. Sci Rep. 2020;10(1):1–7. https://doi.org/10.1038/s41598-020-64630-z.

Tay C, Lindsey EO, Chin ST, McCaughey JW, Bekaert D, Nguyen M, Hua H, Manipon G, Karim M, Horton BP, Li T, Hill EM. Sea-level rise from land subsidence in major coastal cities. Nat Sustain. 2022;5(12):1049–57. https://doi.org/10.1038/s41893-022-00947-z.

The Economist. Say goodbye to 1.5°C. The Economist-Cover, 2022, November 5. https://www.economist.com/weeklyedition/2022-11-05

The Lancet. Health in a world of extreme heat. Lancet. 2021;398(10301):641. https://doi.org/10.1016/S0140-6736(21)01860-2.

Thompson A. Deadly African drought wouldn't have happened without climate change -scientific American Scientific American, 2023, April 27. https://www.scientificamerican.com/article/deadly-african-drought-wouldnt-have-happened-without-climate-change/.

Thomson MC, Stanberry LR. Climate change and Vectorborne diseases. N Engl J Med. 2022;387(21):1969–78. https://doi.org/10.1056/NEJMRA2200092/SUPPL_FILE/NEJMRA2200092_DISCLOSURES.PDF.

UN, Department of Economic and Social Affairs, & Population Division. World Urbanization prospects The 2018 revision, 2019.

UNEP. Adaptation gap report 2022 | UNEP - UN Environment Programme. UN Environment Programme, 2022, November 1. https://www.unep.org/resources/adaptation-gap-report-2022.

UNFCCC. What is the United Nations framework convention on climate change? | UNFCCC, n.d.. United Nations Climate Change. Retrieved July 8, 2023, from https://unfccc.int/process-and-meetings/what-is-the-united-nations-framework-convention-on-climate-change.

UN-Habitat. World cities report 2022 envisaging the future of cities, 2022.

USAID. Environment and climate change | Madagascar | U.S. Agency for International Development. United States Agency for International Development, n.d.. Retrieved July 8, 2023, from https://www.usaid.gov/madagascar/environment-and-climate-change.

Voosen P. Studies tying weather extremes to global warming gain rigor. Science. 2022;376(6599):1256–7. https://doi.org/10.1126/science.add4866.

Wallach PA. Where does US climate policy stand in 2019? Brookings, 2019, March 22. https://www.brookings.edu/articles/where-does-u-s-climate-policy-stand-in-2019/.

Watts N, Amann M, Ayeb-Karlsson S, Belesova K, Bouley T, Boykoff M, Byass P, Cai W, Campbell-Lendrum D, Chambers J, Cox PM, Daly M, Dasandi N, Davies M, Depledge M, Depoux A, Dominguez-Salas P, Drummond P, Ekins P, et al. The Lancet Countdown on health and climate change: from 25 years of inaction to a global transformation for public health. Lancet. 2018;391(10120):581–630. https://doi.org/10.1016/S0140-6736(17)32464-9/ATTACHMENT/EE34AD8C-81B3-472B-A358-F5DF43C849E1/MMC1.PDF.

Whyte C. Running out of river. New Sci. 2022;256(3414):40–3. https://doi.org/10.1016/S0262-4079(22)02136-4.

Wigginton NS, Fahrenkamp-Uppenbrink J, Wible B, Malakoff D. Cities are the future. Science. 2016;352(6288):904–5. https://doi.org/10.1126/SCIENCE.352.6288.904/ASSET/FFA80F4B-D871-436A-B54C-2F2FC7A4CB1B/ASSETS/GRAPHIC/352_904_F1.JPEG.

WMO. State of the global climate 2022. World Meteorological Organization, 2022. https://storymaps.arcgis.com/stories/6d9fcb0709f64904aee371eac09afbdf.

Yao F, Livneh B, Rajagopalan B, Wang J, Crétaux J-F, Wada Y, Berge-Nguyen M. Satellites reveal widespread decline in global lake water storage. Science. 2023;380(6646):743–9. https://doi.org/10.1126/SCIENCE.ABO2812.

Yaro R, Gutman D. Opinion | The Army Corps's flood plan could Ruin New York's Waterfront. It doesn't have to. The New York Times, 2023, June 15. https://www.nytimes.com/interactive/2023/06/15/opinion/nyc-flood-waterfront-plan.html.

2 Climate Change and Health: Geopolitical Perspectives

M. D. Nalapat, Dhanasree Jayaram, and Kiran Bhatt

Abstract

The interlinkages between climate change on the one hand, and security and geopolitics on the other hand, have been altered and redefined in multiple ways, driving the international community to think beyond traditional notions of both security and the phenomenon of climate change itself. Neither geopolitics and security are unaffected by socio-ecological changes; nor can climate or weather be delinked from the changing social, political, economic, cultural, and other factors that are entrenched in life on Earth. This chapter examines public health discourses and practices through a broader lens of human well-being and human security, an assessment that considers the wider systemic causes of the challenges the world faces today, starkly illustrated by the COVID-19 pandemic. It also provides an Indian perspective on various geopolitical and domestic developments and their interconnections with the nexus between climate change and health policies, with an eye on the different stakeholders, disciplines, approaches, and tools required to address the combined and intersecting challenges posed by them.

Keywords

Geopolitics · Security · India · World Health Organization · United Nations Framework Convention on Climate Change

M. D. Nalapat
Department of Geopolitics and International Relations, Manipal Academy of Higher Education, Manipal, India

D. Jayaram (✉)
Department of Geopolitics and International Relations, Manipal Institute of Social Sciences Humanities and Arts, Manipal Academy of Higher Education, Manipal, India
e-mail: dhanasree.j@manipal.edu

K. Bhatt
Centre for Health Diplomacy, Department of Global Health Governance, Prasanna School of Public Health, Manipal Academy of Higher Education, Manipal, India

H. Brand et al. (eds.), *Climate Change and Public Health*,
https://doi.org/10.1007/978-981-95-0895-2_2

2.1 Climate Change, Geopolitics, and Security

The interlinkage between climate change on the one hand, and security and geopolitics on the other hand, have been altered and redefined in multiple ways, driving the international community to think beyond traditional notions of both security and the phenomenon of climate change itself. Neither is geopolitics and security unaffected by socio-ecological changes; nor can climate or weather be delinked from the changing social, political, economic, cultural, and other factors that are being entrenched in life on the Earth. There is, therefore, the need to broaden our bases of inquiry, and thereby our understanding of such changes, as well as to explore the boundaries of various shifts in conditions that the world is undergoing. For instance, environmental degradation and climate change are often regarded as critical causes of increasing poverty and suffering, which in turn are associated with migratory patterns (Hoffmann et al. 2021). Yet, one needs to acknowledge that there are other motivations for communities and individuals to migrate, such as dominant extremist ideologies that persist in certain geographical locations, as seen in the cases of Pakistan and Afghanistan, where Christians, Hindus, and even the Shias and the Ahmadis are forced to migrate due to the persecution and discrimination caused by religious exclusivism (Ispahani 2017). A narrow focus on simply the environmental aspects of climate change would exclude these factors, resulting in insufficiently accurate analysis, with potentially counterproductive and adverse policy implications. One, therefore, needs to view public health through a broader lens of human well-being and human security, an assessment that considers the wider systemic causes of the challenges that the world faces today, which has been starkly illustrated by the COVID-19 pandemic (Zang et al. 2021).

While discussing the factors concerned with responding to climate change, the causal aspects need to be analyzed to design and implement policies. There has been an overhang of negativity concerning its continued use in certain parts of the world due to problems caused by the use and exploitation of fossil fuels for a long time without considering the adverse effects on the climate and the broader environment (Suranovic 2013). The fact is that the populations who are least responsible for greenhouse gas (GHG) emissions face the worst impacts of climate change. The industrialized countries have a historical responsibility for creating the problem by emitting the most amount of GHGs in cumulative and per capita terms (Friman and Strandberg 2014). The worst affected countries, on the other hand, lie in the so-called Global South that lacks the basic resources and capacities to cope with the scale of the problem (Gonzalez and Atapattu 2016). The disagreements between the industrialized and developing countries over climate finance and mitigation continue to derail climate talks (PTI 2023b), particularly with the industrialized countries demanding more ambitious climate action from developing countries with low per capita incomes such as India, while not being willing to pay their dues where climate finance is concerned (Jayaram and Suresh 2023).

Since the establishment of the United Nations Framework Convention on Climate Change (UNFCCC) in the early 1990s, the debates on equity and climate justice have persisted, which led to the operationalization of the principle of Common but Differentiated Responsibilities and Respective Capabilities (CBDR-RC) (Rajamani

2023). Based on the historical responsibility of the industrialized countries (such as the USA, Europe, Canada, and Australia), the developing countries demanded that the former provided technological and financial assistance to the latter to address climate change. Under the Kyoto Protocol (signed in 1997 and entered into force in 2005), a top-down climate agreement, while the industrialized countries were obligated to reduce their emissions, the developing countries including India, were exempted from these obligations. However, countries such as the USA (signed but not ratified the Kyoto Protocol) and Canada (ratified but later withdrew from it) did not fulfill their obligations (Shishlov et al. 2016). Over time, countries such as India and China also adopted voluntary emissions reduction targets (especially after the 2009 Copenhagen Summit), in part due to international pressure (from the industrialized countries), rising GHG emissions, and the aspiration to be recognized as a "responsible power" (Jayaram 2021). India has consistently raised issues related to climate justice such as per capita emissions, equitable sharing of atmospheric (carbon or emissions) and development space, and relative vulnerability which are even today at the center of the climate change negotiations (Khor 2012). While progress on climate action on the part of the industrialized countries is slow, the consequences of gross climate injustice reverberate across the world, particularly in developing countries, with adverse implications for sectors such as public health in these vulnerable regions.

Similarly, the climate change issue is sidetracked in effect because of ongoing geopolitical crises. For example, the continuing Russia-Ukraine War that began in early 2022 substantiates this hypothesis, as the consequent sanctions imposed by some countries on others have caused multiple bottlenecks in the supply chains of various basic goods, including food and energy that are themselves also impacted by climate change. Food security is further threatened as agricultural yields are reduced by droughts, floods, and other impacts of climate change (Wheeler and Von Braun 2013). The European Union's (EU) policy of blocking Russian gas and related economic shocks has led to a continent looking toward a "greener" direction once again investing in coal-based power plants and alternative sources of natural gas and crude oil (Africa, the USA, Middle East, and Central Asia) as emergency measures, which could have devastating impacts on the climate (Meredith 2022). Yet, the worst sufferers of these measures are the underprivileged countries that are not linked to the conflict and are not connected with Russia, Ukraine, or NATO (Ben Hassen and El Bilali 2022; Guan et al. 2023). Thus, there is a need to broaden the idea of climate to refer to humankind and view it as an overall "climate for people" and "climate for societies" across the globe, and not focus only on the privileged populations of the world.

2.2 Geopolitics of Clean Energy Technology and Policy: An Indian Perspective

India has shown a commitment to addressing climate change, particularly by investing in solar, wind, and other forms of clean energy. These choices are also often influenced by geopolitical imperatives. As India is heavily dependent on crude oil imports to meet its energy requirements, there have been discussions on the impact of geopolitical and geoeconomic constraints on energy supplies and the consequent increase in the costs of oil and gas. Consequently, the need for energy independence has led to placing more emphasis on investments in clean energy that could render India more self-reliant (Ramkumar et al. 2021). The geopolitical conundrum of sustainability versus growth often associated with fossil fuels also presents several challenges to India's energy security requirements. While there have been extensive discussions concerning measures to abandon coal, the same is absent in the cases of oil and gas. At the Conference of Parties (COP) that took place in Sharm el-Sheikh in 2022, India raised concerns about singling out coal and demanded the inclusion of the phase-down of all fossil fuels, including oil and gas, in the final text of the agreement reached in the conference (Jayaram 2022).

While many technologies have been developed to harness clean energy, there is a lack of substantial impact on improving the state of the climate. For example, while the scientific, technological, and policy communities have successfully developed and commercialized solar energy systems, short-sighted policies have led to less focus being given to the development of reliable and sustainable energy storage technologies for a long time (Albert and Vanaja 2021). While this critical gap is being bridged in India, there is still a long way to go in terms of achieving energy security through clean energy (Sawhney 2021). Hence, India should lead by strengthening its research and development (R&D) base in energy storage technologies. Similarly, with a greater focus on geopolitics, there has been increased attention in India and elsewhere concerning thorium and its use in nuclear fusion technology to produce clean, accessible, and affordable energy (Garg 2012). It has been argued that if the technology had been developed and used safely and effectively, it could have enhanced benefits for humanity by improving living conditions, including public health, by helping overcome financial barriers to energy access and facilitating the empowerment of the underprivileged (Hamacher et al. 2003). Many of these challenges can be overcome only by addressing the systemic geopolitical problems that curtail the coordination of effective climate action and energy security, as shown by the consequences on inter-country relations caused by the ongoing war in Ukraine.

The abovementioned cases help illustrate the nature of public policy and its failure to improve the living conditions of communities and sustain ecosystems on which human life is dependent. A policy, by its consequences, can be a problem or a solution, and it can be analyzed through its impact on the socio-ecological systems (Hudson et al. 2019). Many of the challenges to the expansion of the solar energy sector in India are attributed to the regulatory systems present within the country. Although there have been several changes in the recent past aimed at increasing domestic manufacturing of solar panels, modules, cells, and other materials required for harnessing solar energy, internal transformation is hindered primarily as it is much cheaper and easier to source them (including raw materials such as critical minerals) from outside (Shiradkar et al. 2022). Moreover, the venture capital movement in India has been slow and has only recently started to gain momentum (S. Garg and Jain 2022). The entire ecology of innovation in India has remained relatively unfriendly for quite some time (Pachouri and Sharma 2016). There has been a trend of young Indians responsible for technological innovation relocating to other countries due to various legal, procedural, administrative, and other obstacles (Mazumdar-Shaw 2020). Here again, geopolitics plays an important role as many of these constraints are linked with aspects related to intergovernmental regimes of international trade, intellectual property rights, and so on. Another factor is the priority given within bureaucratic structures to process over product and control over performance (Nalapat 2014).

What such cases also show is that the world needs to shift away from "silo" approaches. For instance, a combination of the Russia-Ukraine war, rising energy costs, and climate change-related heat waves (in major wheat-producing countries like India) affected global wheat supplies in 2022 (Rennison 2022). It has affected food and energy security, with ripple effects on public health and the economies of several developing and developed countries. It is not enough to consider only one of these issues and come up with policies and solutions designed for it in isolation without considering the question of dealing with the other issues simultaneously. The current mindset or way of functioning where policymaking is concerned is comparable to a "silo" rather than an integrated approach. This has led to several problems in multiple fields, such as in scholarship, bureaucracy, and organizational functioning, each affecting how people deal with their daily lives.

What is crucial because of the nexus between climate change and health is that different stakeholders, disciplines, approaches, and tools are required to address the combined and intersecting challenges posed by them (Machalaba et al. 2015). There needs to be greater coordination between different international organizations dealing with the issues of climate change and health. For instance, the World Health Organization (WHO) has over some time become an integral part of the COP intergovernmental processes on climate change in different capacities, especially for "strengthening joint advocacy with NGOs, UN, academic, and private sector partners on key health and climate cross-cutting issues" (WHO n.d.). The WHO is also known to collaborate with the UNFCCC on the "development of guidelines for assessing the health impacts of climate change," producing "guidance on adaptation measures for human health," and supporting "member states to improve health protection in international health and climate change negotiations and agreements" (WHO 2020). Yet there is a huge gap in providing and implementing concrete solutions (Fox et al. 2019). The UNFCCC has also failed to formally and effectively integrate health concerns into international climate agreements such as the Paris Agreement (Willetts and Haines 2023). The co-benefits of climate action for the public health sector are however increasingly recognized, monitored, and evaluated as a part of climate policy planning and implementation in several countries including India (Kınay and Ji 2022). These developments have not been impervious to geopolitical dynamics. Disputes between countries over how to address global challenges such as climate and health crises have delayed much-needed action at the international level.

2.3 The Future of Climate Action and Public Health: Geopolitical Perspectives

The changing geopolitical realities, particularly with US-China rivalry in the trade, technology, and military domains, potentially provide India an opportunity to establish itself as an alternative renewable energy manufacturing base to China as countries attempt to diversify their supply and value chains and partially decouple from China (Banerjee 2021). India is at present also heavily dependent on China for raw materials and products related to the clean energy sector, forcing the government to introduce fresh policies to reduce the dependence on China. Yet India has found it difficult to reduce imports, despite its geopolitically strained relationship with China (Murray 2023). These geopolitical dynamics have a wide-ranging impact on several sectors, including health, as India's climate change and clean energy policies are deeply intertwined with them. Moreover, the current geopolitical uncertainties have made it increasingly difficult to gauge the future of technology transfer and data sharing between countries, including critical sectors like public health. In the case of China, data sharing on transboundary rivers has not been carried out to the extent necessary where other countries, such as lower riparian states, are concerned (Basist and Williams 2020; Wuthnow 2017).

India has put the issues of climate change and health on its G20 presidency agenda. It strives to build global momentum to advance collective and concerted action on both climate action and public health through its presidency, which is themed—*Vasudhaiva Kutumbakam* (One Earth—One Family—One Future). Focusing on the needs of developing countries, India has pushed for investments in effective and sustainable health systems alongside climate mitigation and adaptation (ADB 2023). The G-20 working group on health discussed the impacts of climate change on health by adopting the "One Health" framework and acknowledging the multidimensional aspects of the linkages between climate change, health, human development, and economics (PTI 2023a). The 2023 G-20 presidency is also an opportunity for India to strengthen its domestic policies on climate change and health. The Asian Development Bank (ADB), for instance, has shown a commitment to establishing a climate change and health hub with the Government of India that will work toward increasing climate finance in healthcare systems (TNN 2023). The future of climate and health policy should encompass an integrated approach that also takes cognizance of the current geopolitical realities. What is essential is a broader context for analyses and consequent policy formulation, a scan that includes social justice, empowerment of marginalized communities, and ecological sustainability.

References

ADB. India's G20 presidency: an opportunity to steer the world toward inclusive and sustainable growth | Asian Development Bank. Asian Development Bank (ADB), 2023, June 19. https://www.adb.org/news/features/indias-g20-presidency-opportunity

Albert JR, Vanaja DS. Solar energy assessments in various regions of Indian sub-continent. In: Elseman AM, editor. Solar cells: theory, materials and recent advances. IntechOpen; 2021.

p. 51–64. https://www.google.com/books/edition/Solar_Cells/_EAAAQBAJ?hl=en&gbpv=1&dq=solar+energy+battery+storage+a+problem+india&pg=PA51&printsec=frontcover.

Banerjee M. Beijing's lead in renewable energy: why India needs to introspect? | Manohar Parrikar Institute for Defence Studies and Analyses. Manohar Parrikar Institute for Defence Studies and Analyses, 2021, November 30. https://idsa.in/issuebrief/beijing-lead-in-renewable-energy-mbanerjee-301121.

Basist A, Williams C. Monitoring the quantity of water flowing through the upper Mekong Basin under natural (unimpeded) conditions, 2020. https://558353b6-da87-4596-a181-b1f20782dd18.filesusr.com/ugd/bae95b_0e0f87104dc8482b99ec91601d853122.pdf.

Ben Hassen T, El Bilali H. Impacts of the Russia-Ukraine War on global food security: towards more sustainable and resilient food systems? Foods. 2022;11(15) https://doi.org/10.3390/FOODS11152301.

Fox M, Zuidema C, Bauman B, Burke T, Sheehan M. Integrating public health into climate change policy and planning: state of practice update. Int J Environ Res Public Health. 2019;16:3232. https://doi.org/10.3390/IJERPH16183232.

Friman M, Strandberg G. Historical responsibility for climate change: science and the science–policy interface. Wiley Interdiscip Rev Clim Chang. 2014;5(3):297–316. https://doi.org/10.1002/WCC.270.

Garg P. Energy scenario and vision 2020 in India. J Sustain Energy Environ. 2012;3:7–17.

Garg S, Jain R. Making India a leader in solar manufacturing: ways to achieve technology leadership and global competitiveness. Council on Energy, Environment and Water (CEEW), 2022. https://www.ceew.in/cef/solutions-factory/publications/CEEW-Making-India-a-Leader-in-Solar-Manufacturing.pdf.

Gonzalez CG, Atapattu S. International Environmental Law, environmental justice, and the global south. Trans Law Contemp Probl. 2016;26. https://heinonline.org/HOL/Page?handle=hein.journals/tlcp26&id=247&div=&collection=

Guan Y, Yan J, Shan Y, Zhou Y, Hang Y, Li R, Liu Y, Liu B, Nie Q, Bruckner B, Feng K, Hubacek K. Burden of the global energy price crisis on households. Nat Energy. 2023;8(3):304–16. https://doi.org/10.1038/s41560-023-01209-8.

Hamacher T, Shukla RP, Seebregts AJ. The possible role of fusion in the Indian energy system of the future. Fusion Eng Des. 2003;69(1–4):733–7. https://doi.org/10.1016/S0920-3796(03)00117-0.

Hoffmann R, Šedová B, Vinke K. Improving the evidence base: a methodological review of the quantitative climate migration literature. Glob Environ Chang. 2021;71:102367. https://doi.org/10.1016/J.GLOENVCHA.2021.102367.

Hudson B, Hunter D, Peckham S. Policy failure and the policy-implementation gap: can policy support programs help? 2019;2(1):1–14. https://doi.org/10.1080/25741292.2018.1540378.

Ispahani F. Pakistan's descent into religious intolerance | Hudson. Hudson, 2017, March 1. https://www.hudson.org/national-security-defense/pakistan-s-descent-into-religious-intolerance.

Jayaram D. Climate diplomacy and emerging economies: India as a case study. 1st ed. Routledge; 2021. https://www.routledge.com/Climate-Diplomacy-and-Emerging-Economies-India-as-a-Case-Study/Jayaram/p/book/9780367634049

Jayaram D. At COP-27, climate change mitigation takes a back seat | The India Forum. The India Forum, 2022, December 20. https://www.theindiaforum.in/amp/climate-change/failure-cop-27-climate-change-mitigation-takes-back-seat.

Jayaram D, Suresh M. COP27 has reopened the Pandora's Box on 'fairness in climate finance.' 9DASHLINE, 2023, March 6. https://www.9dashline.com/article/cop27-has-reopened-the-pandoras-box-on-fairness-in-climate-finance.

Khor M. The equitable sharing of atmospheric and development space: summary. In: UNFCCC; 2012. https://unfccc.int/files/bodies/awg-lca/application/pdf/20120516_south_centre_paper_1701.pdf.

Kınay P, Ji JS. Reported evidence of greenness co-benefits on health, climate change mitigation, and adaptation: a systematic review of the literature. Environ Res Clim. 2022;1(1):012002. https://doi.org/10.1088/2752-5295/AC4DA2.

Machalaba C, Romanelli C, Stoett P, Baum SE, Bouley TA, Daszak P, Karesh WB. Climate change and health: transcending silos to find solutions. Ann Glob Health. 2015;81(3):445–58. https://doi.org/10.1016/J.AOGH.2015.08.002.

Mazumdar-Shaw K. The challenges and opportunities of innovating in India | ORF. Observer Research Foundation, 2020, October 30. https://www.orfonline.org/expert-speak/challenges-opportunities-innovating-india/.

Meredith S. Russia is squeezing Europe's gas supplies, sparking a bitter and reluctant return to coal. CNBC, 2022, June 21. https://www.cnbc.com/2022/06/21/ukraine-war-europe-turns-to-coal-as-russia-squeezes-gas-supplies.html.

Murray B. Supply chain latest: India decoupling from China - Bloomberg. Bloomberg, 2023, June 8. https://www.bloomberg.com/news/newsletters/2023-06-08/supply-chain-latest-india-decoupling-from-china.

Nalapat MD. The practice of geopolitics | exotic India art. Manipal Universal Press, 2014. https://www.exoticindiaart.com/book/details/practice-of-geopolitics-naz832/.

Pachouri A, Sharma S. Barriers to innovation in Indian small and medium-sized enterprises. Asian Development Bank Institute. 2016;588:1–30. https://doi.org/10.22/JS/JQUERY.DATATABLES.MIN.JS

PTI. G20 Health Working Group discusses impacts of climate change. Outlook Planet, 2023a, June 6. https://planet.outlookindia.com/news/g20-health-working-group-discusses-impacts-of-climate-change%2D%2Dnews-415450

PTI. Bonn climate talks end without concrete conclusions on finance. Outlook Planet, 2023b, June 16. https://planet.outlookindia.com/news/bonn-climate-talks-end-without-concrete-conclusions-on-finance-news-415510.

Rajamani L. Common but differentiated responsibilities. In: Faure M, editor. Elgar Encyclopedia of environmental law. Edward Elgar Publishing; 2023. p. 291–302. https://doi.org/10.4337/9781785369520.

Ramkumar M, Santosh M, Mathew MJ, Siddiqui NA. India at crossroads for energy. Geosci Front. 2021;12(6):100901. https://doi.org/10.1016/J.GSF.2019.10.006.

Rennison J. War, climate change, energy costs: how the wheat market has been upended - Forbes India. Forbes India, 2022, August 1. https://www.forbesindia.com/article/news/war-climate-change-energy-costs-how-the-wheat-market-has-been-upended/78597/1

Sawhney A. Striving towards a circular economy: climate policy and renewable energy in India. Clean Techn Environ Policy. 2021;23(2):491–9. https://doi.org/10.1007/S10098-020-01935-7/TABLES/4.

Shiradkar N, Arya R, Chaubal A, Deshmukh K, Ghosh P, Kottantharayil A, Kumar S, Vasi J. Recent developments in solar manufacturing in India. Solar Compass. 2022;1:100009. https://doi.org/10.1016/J.SOLCOM.2022.100009.

Shishlov I, Morel R, Bellassen V. Compliance of the parties to the Kyoto Protocol in the first commitment period. Clim Pol. 2016;16(6):768–82. https://doi.org/10.1080/14693062.2016.1164658.

Suranovic S. Fossil fuel addiction and the implications for climate change policy. Glob Environ Chang. 2013;23(3):598–608. https://doi.org/10.1016/J.GLOENVCHA.2013.02.006.

TNN. ADB to create climate change and health hub | Goa News - Times of India. The Times of India, 2023, April 20. https://timesofindia.indiatimes.com/city/goa/adb-to-create-climate-change-and-health-hub/articleshow/99649298.cms?from=mdr

Wheeler T, Von Braun J. Climate change impacts on global food security. Science. 2013;341(6145):508–13. https://doi.org/10.1126/SCIENCE.1239402.

WHO. Promoting and protecting health through the UNFCCC. World Health Organization, 2020, January 1. https://www.who.int/news/item/01-01-2020-promoting-and-protecting-health-through-the-unfccc.

WHO. Climate change and health. World Health Organization, n.d.. Retrieved July 31, 2023, from https://www.who.int/teams/environment-climate-change-and-health/climate-change-and-health/advocacy-partnerships/talks/cop27

Willetts E, Haines A. Managing the health risks of climate change. Council on Foreign Relations, 2023, March 6. https://www.cfr.org/report/managing-health-risks-climate-change.

Wuthnow J. Did China use water as a weapon in the Doklam standoff?. War on the Rocks, 2017, October 4. https://warontherocks.com/2017/10/did-china-use-water-as-a-weapon-in-the-doklam-standoff/

Zang SM, Benjenk I, Breakey S, Pusey-Reid E, Nicholas PK. The intersection of climate change with the era of COVID-19. Public Health Nurs. 2021;38(2):321–35. https://doi.org/10.1111/PHN.12866.

Particulate Pollution Around the World and Its Health Implications

3

Imad El Haddad, Kaspar Daellenbach, Jay Slowik, Lubna Dada, Benjamin Brem, and Andre S. H. Prevot

Abstract

Aerosols, tiny solid or liquid particles in the atmosphere, influence climate by absorbing and reflecting sunlight and modifying cloud properties. Light-coloured aerosols, such as sulphate and sea salt, cool the atmosphere, whereas dark aerosols, such as black carbon, cause warming. These interactions also affect cloud formation and properties. Unlike greenhouse gases, aerosols have a net cooling effect; however, their impact on Earth's radiative budget remains uncertain in quantifying human contributions to global warming. Additionally, aerosols pose health risks, emphasizing the need to understand their sources, formation pathways, and chemical composition. This chapter examines the impacts of aerosol particles on Earth's climate and human health and reviews ambient measurements of aerosol sources across different geographies, focusing on the significance of extensive observations and modelling efforts for addressing climate change and improving air quality.

Keywords

Aerosols · Climate change · Human health · Atmospheric cooling · Cloud formation · Radiative budget · Air quality

I. El Haddad (✉)
Group Head Molecular Cluster and Particle Processes, Paul Scherrer Institute, Wurenlingen, Switzerland
e-mail: imad.el-haddad@psi.ch

K. Daellenbach · J. Slowik · L. Dada · B. Brem
Laboratory of Atmospheric Chemistry, Paul Scherrer Institute, Wurenlingen, Switzerland

A. S. H. Prevot
Research Group Head, Paul Scherrer Institute, Wurenlingen, Switzerland

H. Brand et al. (eds.), *Climate Change and Public Health*,
https://doi.org/10.1007/978-981-95-0895-2_3

3.1 Introduction

Aerosol particles, which are tiny solid or liquid particles suspended in the atmosphere, have a significant impact on Earth's climate and human health (University of Leeds 2023). Understanding the effects of aerosols is crucial in quantifying the human contribution to global warming and addressing public health concerns.

Aerosols affect our climate both directly, by absorbing and reflecting sunlight, and indirectly, by modifying cloud properties. Some aerosols are light-coloured such as sulphate from industrial emissions or sea salt from ocean spray and have a cooling effect on the atmosphere by scattering and reflecting sunlight. Conversely, other aerosols like black carbon have the opposite effect and can absorb sunlight, leading to atmospheric warming. The complex interactions between aerosols and clouds also play a significant role in the climate system. Aerosols can act as cloud condensation nuclei (CCN) or ice nuclei (IN), affecting cloud formation and properties. CCN provides surfaces on which water vapour can condense, leading to the formation of clouds. By increasing the number of CCN, aerosols can enhance cloud formation and potentially increase cloud lifetime (Penner Lead et al. 2018). Furthermore, aerosols can influence the cloud's brightness and extent, affecting its reflectivity and ability to absorb or reflect sunlight. Unlike GHGs, aerosol particles have a net cooling effect on our atmosphere, yet their impact on Earth's radiative budget[1] remains the single largest uncertainty in quantifying the human contribution to global warming (Fu 2015).

When it comes to human health, aerosols pose significant risks. Fine particulate matter (PM2.5), which includes aerosol particles with a diameter of 2.5 micrometres or less, can penetrate deep into the respiratory system when inhaled. Exposure to PM2.5 is linked to a wide range of health problems, including respiratory and cardiovascular diseases, reduced lung function, and increased mortality rates. It is estimated that more than 7 million premature deaths occur each year due to air pollution, with aerosol particles being a major component (Achanzar and Mangipudy 2014; Manisalidis et al. 2020).

The sources of aerosols are diverse and include natural processes such as volcanic eruptions, dust storms, and sea spray, as well as human activities like industrial emissions, vehicle exhaust, and biomass burning. Aerosol particles can be directly emitted into the atmosphere or formed from low-volatility oxidation products from precursors (Myhre et al. 2013).

Enhancing our understanding of aerosol sources, formation pathways, and chemical composition is vital for accurately quantifying their effects on climate and public health. Furthermore, improving our knowledge of aerosols will help us address climate change and improve air quality for the benefit of both the environment and human health. For a comprehensive understanding of aerosols, a combination of

[1] Radiative forcing is a key concept in understanding the role of aerosols in climate change. It refers to the difference between the amount of radiant energy received by the Earth and the amount of energy radiated back into space at the top of atmosphere

field observations, long-term measurements, chamber experiments, and modelling efforts is needed.

The Laboratory of Atmospheric Chemistry (LAC), within the Energy and Environment Division at Paul Scherrer Institute in Switzerland, runs and maintains comprehensive air quality monitoring stations, which are part of international consortia such as ACTRIS[2] and GAW.[3] These include the Jungfraujoch station, a mountaintop station located at 3580 m a.s.l. within the Swiss Alps (Fig. 3.1) and Payerne station which is a rural background measurement station located in Southwestern Switzerland. Among many climate parameters that are measured at these stations, long-term particle concentration and their optical properties are measured, useful for designing mitigation strategies on climate and air quality. Occasionally, short-term measurement campaigns with extended additional instrumentation, e.g. mass spectrometers, are deployed at either location for a detailed understanding of atmospheric observations. Measurement campaigns run by the LAC are, however, not limited to Jungfraujoch and Payerne. Instead, the LAC has performed measurement campaigns in hundreds of locations, in Europe, East Asia, the Arctic, and the Antarctic. During such campaigns, the instrumentation is transported to the measurement location, where it is deployed for a while, depending on the feasibility and purpose of the measurement campaign.

The LAC also studies air pollution by analysing filter archives in the laboratory (Daellenbach et al. 2016). The method entails nebulizing water extracts into the mass spectrometer for the determination of their bulk and molecular chemical composition. Data factorization techniques are applied to disentangle the different aerosol sources based on their composition (Daellenbach et al. 2016). This technique, despite its limited time resolution, enables the simultaneous measurements of filters within dense monitoring networks over extended periods, access to locations with limited infrastructures (e.g. the Arctic), and the investigations of severe pollution events (Vogel et al. 2017). This technique has enabled the analysis of more than 15,000 samples from more than 50 locations worldwide. Currently, two large projects in the world's most polluted countries: China and India are taking place, funded by the Swiss Bureau

[2]ACTRIS: The aerosol, cloud and trace-gases research infrastructure is a pan-European research infrastructure producing high-quality data and information on short-lived atmospheric constituents and on the processes leading to the variability of these constituents in natural and controlled atmospheres.

[3]GAW: The Global Atmosphere Watch Programme of the World Meteorological Organization (WMO) focuses on building a single coordinated global understanding of atmospheric composition, its change, and helps to improve the understanding of interactions between the atmosphere, the oceans and the biosphere. It coordinates high-quality atmospheric composition observations across global to local scales to drive high-quality and impact science while co-producing a new generation of research-enabled products and services. About 100 countries are participating in the GAW Program. One major aspect of the GAW mission is to organize, participate in and coordinate assessments of the chemical composition of the atmosphere on a global scale. In this way, GAW provides reliable scientific information for policymakers, supports international conventions, and contributes to improving the understanding of climate change and long-range transboundary air pollution.

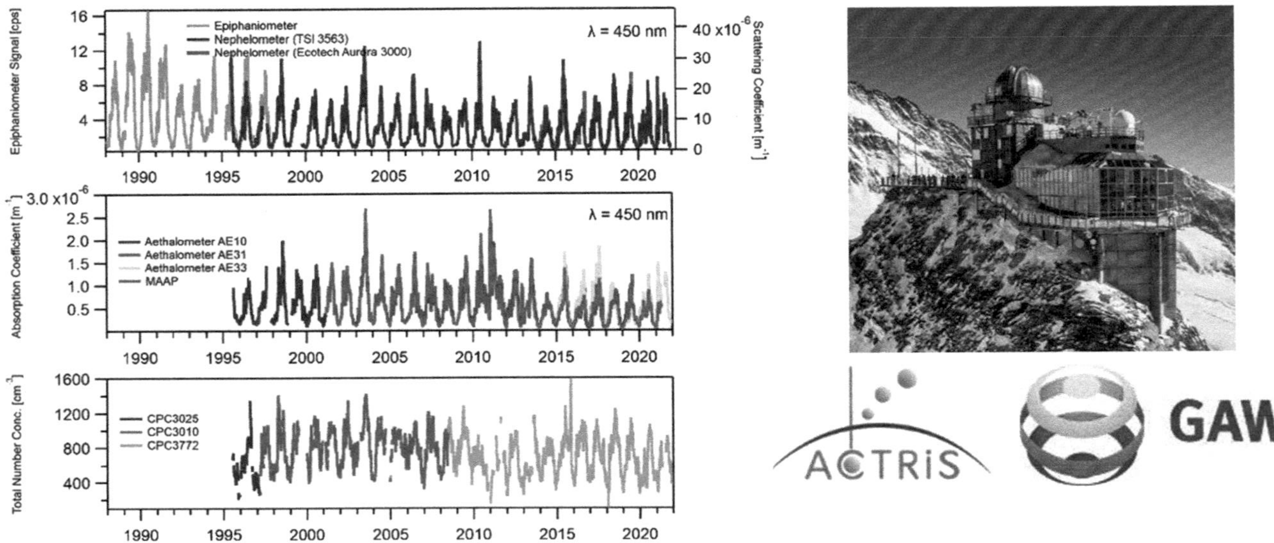

Fig. 3.1 One of the longest-time series of aerosol properties in the world measured at the top of Europe, at the Swiss alpine station of Jungfraujoch. Shown in the figure are the measurements of aerosol scattering, absorption, and number concentration. A picture of the Jungfraujoch station, which is part of ACTRIS and GAW programs, is shown on the right. (More info can be found at (Bukowiecki et al. 2016). Digital Object Repository at PSI. Source: Figure updated and adapted from Bukowiecki et al. 2015, with the support of the Global Atmosphere Watch Program, ACTRIS Switzerland, and the High-Altitude Research Stations Jungfraujoch and Gornergrat. Photograph of the Jungfraujoch Sphinx research is provided in courtesy of the Jungfrau Railways)

for Development and Cooperation.[4] Filters from multiple locations in both countries over several months have been collected and analysed in our laboratories.

The LAC is a steering member in two eminent consortia: ACTRIS and CLOUD, which practically encompass all the expertise in atmospheric simulation chambers in Europe. We operate and provide access to our unique chamber facility at PSI and perform chamber experiments at CERN to study new particle formation. Chamber experiments act as controlled settings for performing experiments that simulate different chemical processes of the atmosphere. For example, wood-burning experiments are performed at PSI with different types of wood, to generate chemical fingerprints, which can later be compared to ambient observations. The parameters derived from chamber experiments are used to improve models, with which we assess the air quality on regional scales and the impacts of different aerosol components on public health.

In this chapter, we will discuss ambient measurements we have performed to characterize the sources of aerosols in different geographies. A detailed understanding of the sources and the nature of aerosols is important to identify appropriate measures to manage the health impacts of aerosols.

3.2 Long-Term Climate Variables at Jungfraujoch

The measurements of more than 30 years on top of the Jungfraujoch allow for an understanding of long-term trends of air pollutants and the effectiveness of previously proposed mitigation strategies. Based on the collected data, here particle concentrations, and scattering/absorbing coefficients, we observed a decrease in absorbing and scattering aerosols since the early 2000s due to sulphur removal from fuels and reduced soot emissions in the western hemisphere (Collaud Coen et al. 2020).

3.3 Overview of Aerosol Sources Across Europe

Besides studying short-lived climate variables at the Jungfraujoch site, we work on the harmonization of measurements performed at multiple locations in Europe as part of the ACTRIS program. In a recent study led by the LAC, an overview of the aerosol chemical composition and sources is derived (Chen et al. 2022). The results show that in Europe, whether it is a site that is polluted like in Poland or sites without pollution, for example, in Sweden or Norway, organics dominate the chemical composition of the aerosol (Fig. 3.2). These organics are mostly of secondary origin, resulting from reactions of organic precursors in the atmosphere. In winter, biomass burning for domestic heating dominates direct emissions of aerosols. During the same season, ammonium nitrate, from the oxidation of mobile emissions of NO_X, can also be an important fraction at many sites (Chen et al. 2022).

[4] Responsible for the overall coordination of development activities and cooperation with Eastern Europe, as well as for the humanitarian aid delivered by the Swiss Confederation.

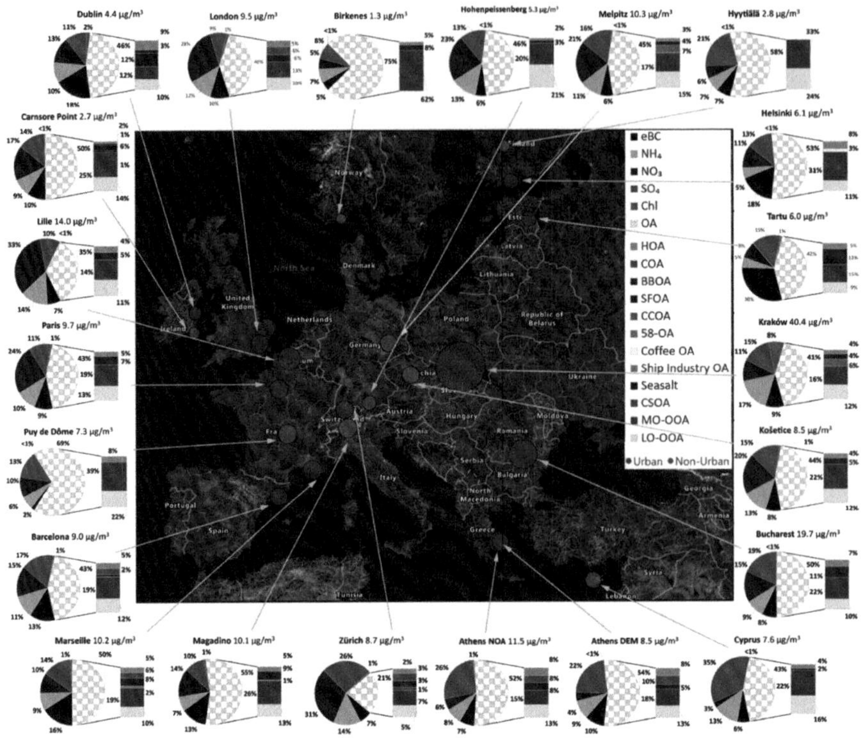

Fig. 3.2 European map of aerosol composition and sources. (By Chen et al. 2022)

3.4 Particulate Air Pollution Sources in Polluted Megacities

One of the first applications of the offline filter analysis technique (Daellenbach et al. 2016) was the characterization of aerosol sources during record-breaking pollution in January 2013 event with $PM_{2.5}$ exceeding 800 μg/m^3 (Huang et al. 2014). For this, we have analysed filter samples from Chinese megacities: Beijing, X'ian, Shanghai, and Guangzhou. Our analysis showed that the oxidation of vapours in the atmosphere is the main driver of aerosol pollution and haze formation, which was the first discovery of its kind. This aerosol fraction is termed secondary. Before our study, the common understanding was that oxidation during winter was too slow and that direct emission of particles is the main driver of pollution. Our results imply that limiting direct emissions of particles, e.g. using particle filters, will not be the most effective way to limit haze formation in China. We have also determined the main sources of primary aerosols, including biomass and coal burning as well as desert dust in certain cities, especially those in proximity to the Gobi Desert (Huang et al. 2014).

In later years, we investigated the composition and the sources of the dominant secondary fraction using near-molecular analysis. We could identify that during haze, the secondary fraction is produced from the ageing of biomass burning emissions and aqueous phase processes in the particle phase at high relative humidity (Tong et al. 2021). Aqueous phase processes appear to be promoted by the presence of nitrate particles from anthropogenic emissions, which increases the particle water content offering a medium for additional reactions. Overall, the main source of secondary aerosol precursors is residential burning; however, traffic emissions of NO_X indirectly promote SOA formation through particle phase processing (Tong et al. 2021).

The situation in India, especially in New Delhi, is different with substantially higher pollution levels from local primary emissions. Compared to China, where average wintertime $PM_{2.5}$ concentrations are ~70 μg/m³ (Yan et al. 2022), the concentrations in New Delhi exceed a daily average of 350 μg/m³ (Rai et al. 2020). On some days, the concentrations approach 1000 μg/m³. As a comparison, the concentrations in Zurich, for example, are ~10 μg/m³. Interestingly, Delhi experiences very high loadings of ammonium chloride (Mishra et al. 2023). Particulate chloride is formed from the condensation of hydrochloric acid, which is enhanced during night-time at lower temperatures and high relative humidity. This process has not been found anywhere else in the world. The sources of hydrochloric acid remain unknown in Delhi, but higher particulate chloride concentrations have been observed when air masses come from the industrial areas around New Delhi (Mishra et al. 2023). The organic aerosol dominates the total PM mass in Delhi relative (Fig. 3.3) to the high contribution of primary emissions from vehicular exhausts and biomass burning (Tobler et al. 2020). The secondary fraction dominates during the daytime (Kumar et al. 2022), due to the oxidation of traffic and biomass-burning emissions. Overall, limiting incomplete combustion sources in India will be extremely beneficial for both air quality and human health.

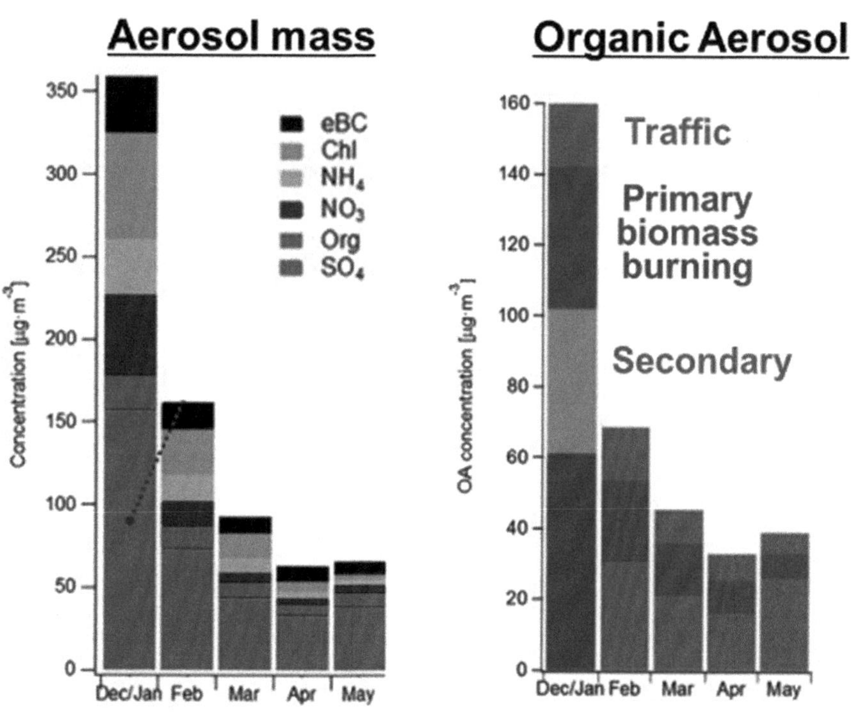

Fig. 3.3 Aerosol composition and organic aerosol sources in Delhi India. (By Tobler et al. 2020)

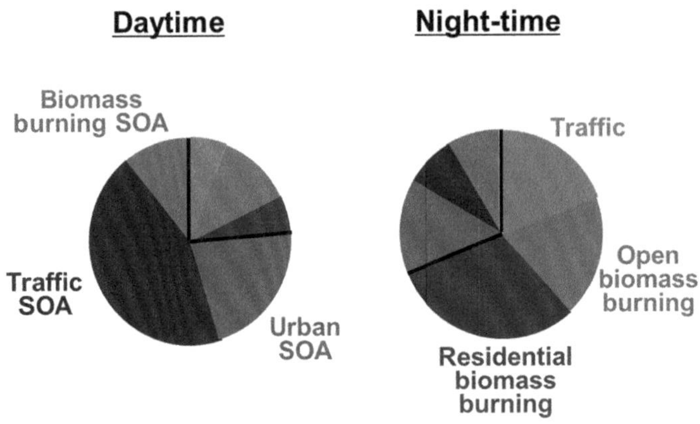

Fig. 3.4 Detailed apportionment of the organic aerosol sources during daytime and nighttime, using molecular-level information. (Adapted from Kumar et al. 2022. Copernicus)

3.5 Sources of Particulate Air Pollution in the Arctic

Pollution travels far and reaches pristine areas. We have investigated the aerosol composition in the Arctic, a region that is warming three times faster than the rest of the globe (Dada et al. 2022; Rantanen et al. 2022), where aerosols play a major role by forming low-level clouds. This warming will increase the surface of open oceans, allowing for additional vapour emissions from the phytoplankton, which may in turn lead to further aerosol formation and growth, creating a positive feedback loop and accelerating the arctic amplification.[5] The sources of the organic aerosol fraction arriving in different sectors in the Arctic remain unknown, especially over the long term. In a recent study, we applied the offline technique described above to 350 filter samples collected over the past ten years covering eight sites in six countries within the Arctic Circle (Moschos et al. 2022). Our findings show that in winter the organic aerosol that forms the arctic haze is dominated by anthropogenic emissions because of direct combustion as well as long-range transported aged aerosols from Eurasia. By contrast, natural emissions dominate the organic aerosol composition during summer. These include marine secondary emissions from the Arctic Ocean, together with primary and secondary natural emissions transported from the Siberian boreal biome. Our quantification revealed equal contributions to arctic organic aerosol from natural and anthropogenic emissions. Future studies can use our data to constrain models estimating the role of aerosols in the changing Arctic climate (Fig. 3.4).

[5] Arctic amplification refers to the phenomenon where the Arctic region is warming at a faster rate than the rest of the planet. This is primarily driven by positive feedback mechanisms. As the Arctic Sea ice melts due to rising temperatures, the exposed dark ocean absorbs more sunlight, leading to further warming and ice loss, creating a feedback loop that intensifies the warming trend and has wide-ranging impacts on global climate patterns.

3.6 Chamber Experiments

We use laboratory simulation experiments to better understand our atmosphere and derive simple parameterization that can be used in models to represent different aerosol sources and formation processes. During these experiments, we study the chemical composition and the ageing of complex emissions from biomass burning, cooking, and fossil fuel combustion. More recently, we have investigated the solid fuel combustion processes that take place in India, including the burning of waste and cow dung.

Our results show that emissions of primary organic particles are important from biomass burning, while 95 percent of biomass smoke is composed of volatile vapours. Among these, the most important for secondary aerosol formation are furans from cellulose pyrolysis, oxygenated aromatic compounds from lignin pyrolysis, and poly-aromatic compounds from incomplete combustion. The secondary organic aerosol fraction formed from these precursors exceeds the primary fraction after only a few hours of oxidation under atmospheric conditions. Our results provide a mechanistic understanding of the ageing processes by which major emissions transform in the atmosphere.

3.7 Modelling Aerosol Composition and Sources in Europe

The parameters derived from chamber experiments are used to feed models which allow investigation of the sources of particulate air pollution on a large scale. We compared our modelling results to measurements across Europe (Fig. 3.5), showing the accuracy of the model in capturing the site-to-site differences, temporal variability as well as chemical composition of the aerosols. The model is also capable of differentiating the sources of aerosols, showing that 30 percent of the organic fraction could be attributed to primary emissions from traffic and biomass burning, while the rest is secondary from the oxidation of biomass burning emissions in winter and biogenic emissions in summer (Fig. 3.6). With this, we have a full picture of the composition and the sources of aerosol across Europe. Similar modelling analysis can be applied to more polluted locations like India and China to discover the most important sources of aerosol pollution.

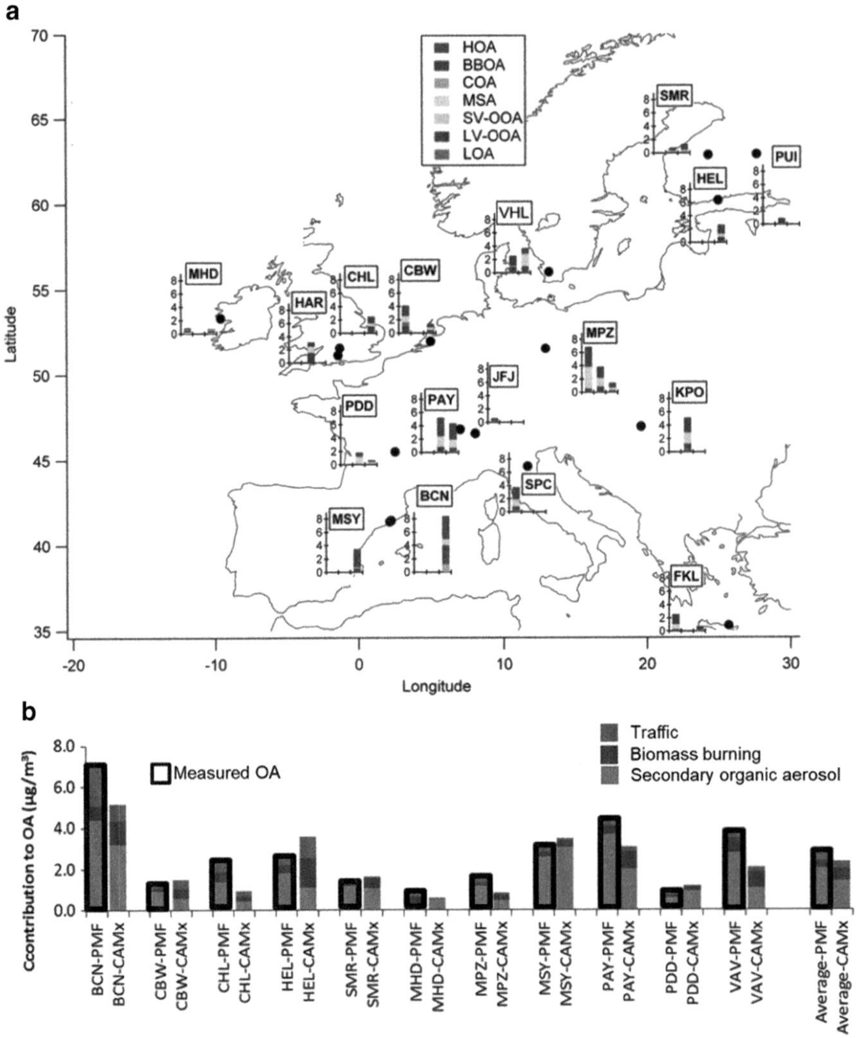

Fig. 3.5 Comparison of modelling and measurements of the organic aerosol across Europe. (Adapted from Ciarelli et al. 2017. Copernicus)

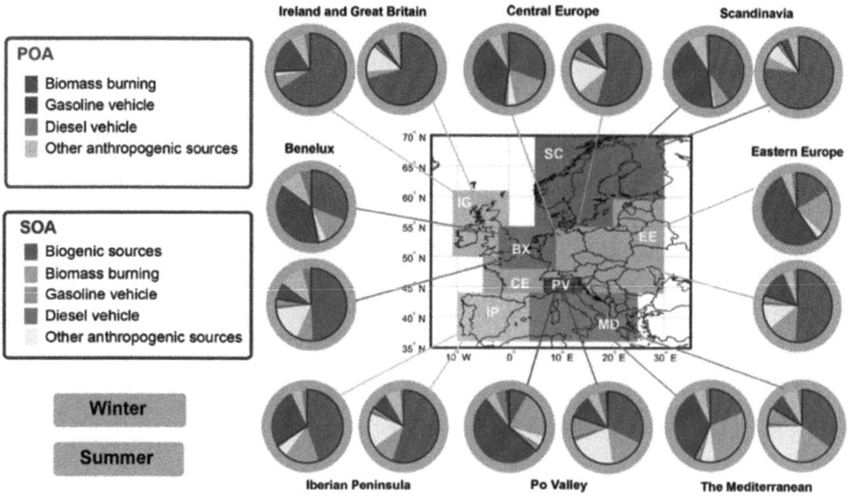

Fig. 3.6 Organic aerosol sources modelled across Europe. (By Jiang et al. 2019)

3.8 Health Impacts of Atmospheric Aerosols

Aerosol health effects are based on the total mass concentrations, while its toxicity is dependent on its chemical composition and hence its sources. In a recent study (Daellenbach et al. 2020), we investigated whether the same sources drive the total aerosol mass and its oxidative potential (OP),[6] which is thought to lead to acute health effects. Via combined measurement and modelling efforts, maps of OP sources are derived, showing hotspots of aerosol toxicity, where high pollution meets high population in Europe, e.g. Paris and Madrid. Across Europe, the overwhelming sources of aerosol mass are found to be secondary inorganic aerosols and crustal material from desert and road dust, while the sources of OP are different, dominated by anthropogenic secondary organic aerosols and vehicular wear (Fig. 3.7) (Daellenbach et al. 2020). Our results imply that focusing on the reduction of the total aerosol mass would not result in an effective reduction of aerosol health effects.

The chemical analysis of aerosols, combined with interdisciplinary collaborations, plays a crucial role in our understanding of air pollution and its impact on human health. By characterizing aerosols at various locations, we provide valuable data that empowers health professionals and policymakers to make informed decisions regarding pollution management and regulation of pollutants. This comprehensive approach helps address the individual and public health costs associated with pollution. Our ongoing work, in collaboration with epidemiologists, data scientists, and medical doctors, allows us to model aerosol sources and assess

[6] Oxidative potential (OP) is a measure of the aerosol content of reactive oxygen species which reacts with anti-oxidants in our body.

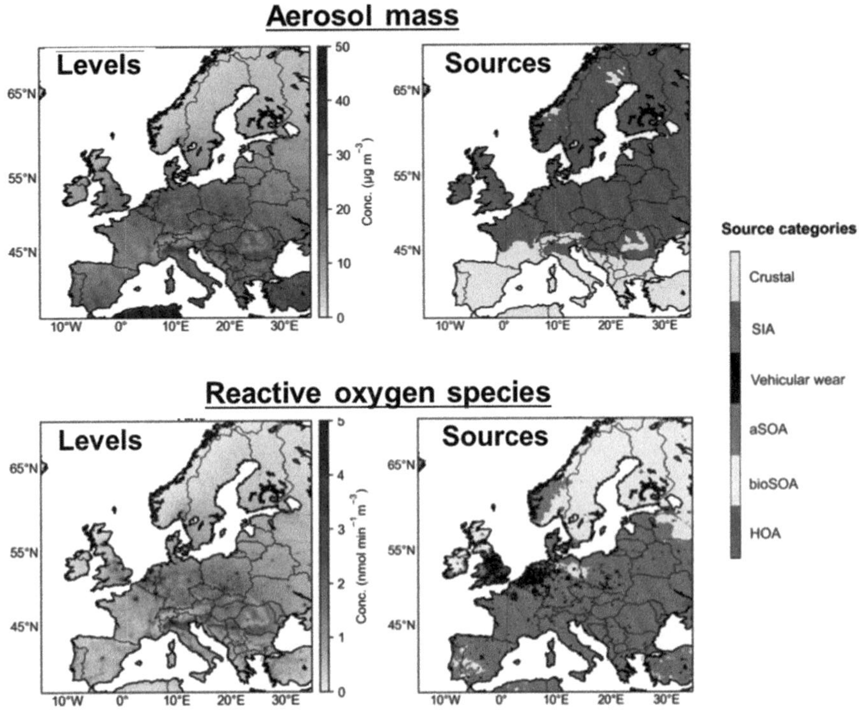

Fig. 3.7 Comparison between the sources of aerosols and reactive oxygen species. (By Daellenbach et al. 2020)

personalized exposures to specific aerosol components. Through extensive atmospheric observations and the collection of over 5,00,000 data points on aerosol composition across Europe, we are building models that capture personalized exposures to aerosol sources and their components. This interdisciplinary understanding, along with direct engagement with the public, enables us to identify the most harmful particulate matter components that pose significant risks to our health. Together, these efforts contribute to a more comprehensive and effective approach to managing air pollution and safeguarding public health.

References

Achanzar WE, Mangipudy RS. Aerosols. In: Encyclopedia of toxicology. 3rd ed; 2014. p. 100–101. https://doi.org/10.1016/B978-0-12-386454-3.00223-2.

Bukowiecki N, Weingartner E, Gysel M, Coen MC, Zieger P, Herrmann E, Steinbacher M, Gäggeler HW, Baltensperger U. A review of more than 20 years of aerosol observation at the high altitude research station Jungfraujoch, Switzerland (3580 m asl). Aerosol Air Qual Res. 2016;16(3):764–88. https://doi.org/10.4209/AAQR.2015.05.0305.

Chen G, Canonaco F, Tobler A, Aas W, Alastuey A, Allan J, Atabakhsh S, Aurela M, Baltensperger U, Bougiatioti A, De Brito JF, Ceburnis D, Chazeau B, Chebaicheb H, Daellenbach KR,

Ehn M, El Haddad I, Eleftheriadis K, Favez O, et al. European aerosol phenomenology – 8: Harmonised source apportionment of organic aerosol using 22 Year-long ACSM/AMS datasets. Environ Int. 2022;166:1–18. https://doi.org/10.1016/J.ENVINT.2022.107325.

Ciarelli G, Aksoyoglu S, El Haddad I, Bruns EA, Crippa M, Poulain L, Äijälä M, Carbone S, Freney E, O'Dowd C, Baltensperger U, Prévôt ASH. Modelling winter organic aerosol at the European scale with CAMx: Evaluation and source apportionment with a VBS parameterization based on novel wood burning smog chamber experiments. Atmos Chem Phys. 2017;17(12):7653–69. https://doi.org/10.5194/ACP-17-7653-2017.

Collaud Coen M, Andrews E, Lastuey A, Petkov Arsov T, Backman J, Brem BT, Bukowiecki N, Couret C, Eleftheriadis K, Flentje H, Fiebig M, Gysel-Beer M, Hand JL, Hoffer A, Hooda R, Hueglin C, Joubert W, Keywood M, Eun Kim J, et al. Multidecadal trend analysis of in situ aerosol radiative properties around the world. Atmos Chem Phys. 2020;20(14):8867–908. https://doi.org/10.5194/ACP-20-8867-2020.

Dada L, Angot H, Beck I, Baccarini A, Quéléver LLJ, Boyer M, Laurila T, Brasseur Z, Jozef G, de Boer G, Shupe MD, Henning S, Bucci S, Dütsch M, Stohl A, Petäjä T, Daellenbach KR, Jokinen T, Schmale J. A central arctic extreme aerosol event triggered by a warm air-mass intrusion. Nat Commun. 2022;13(1):1–15. https://doi.org/10.1038/s41467-022-32872-2.

Daellenbach KR, Bozzetti C, Křepelová A, Canonaco F, Wolf R, Zotter P, Fermo P, Crippa M, Slowik JG, Sosedova Y, Zhang Y, Huang R-J, Poulain L, Szidat S, Baltensperger U, El Haddad I, Prévôt ASH. Characterization and source apportionment of organic aerosol using offline aerosol mass spectrometry. Atmos Meas Tech. 2016;9:23–39. https://doi.org/10.5194/amt-9-23-2016.

Daellenbach KR, Uzu G, Jiang J, Cassagnes LE, Leni Z, Vlachou A, Stefenelli G, Canonaco F, Weber S, Segers A, Kuenen JJP, Schaap M, Favez O, Albinet A, Aksoyoglu S, Dommen J, Baltensperger U, Geiser M, El Haddad I, et al. Sources of particulate-matter air pollution and its oxidative potential in Europe. Nature. 2020;587(7834):414–9. https://doi.org/10.1038/s41586-020-2902-8.

Fu Q. Radiation transfer in the atmosphere | Radiation, Solar. Encyclopedia of Atmospheric Sciences. 2nd ed; 2015. p. 1–4. https://doi.org/10.1016/B978-0-12-382225-3.00334-0.

Huang RJ, Zhang Y, Bozzetti C, Ho KF, Cao JJ, Han Y, Daellenbach KR, Slowik JG, Platt SM, Canonaco F, Zotter P, Wolf R, Pieber SM, Bruns EA, Crippa M, Ciarelli G, Piazzalunga A, Schwikowski M, Abbaszade G, et al. High secondary aerosol contribution to particulate pollution during haze events in China. Nature. 2014;514(7521):218–22. https://doi.org/10.1038/nature13774.

Jiang J, Aksoyoglu S, El-Haddad I, Ciarelli G, Denier Van Der Gon HAC, Canonaco F, Gilardoni S, Paglione M, Minguillón MC, Favez O, Zhang Y, Marchand N, Hao L, Virtanen A, Florou K, O'Dowd C, Ovadnevaite J, Baltensperger U, Prévôt ASH. Sources of organic aerosols in Europe: a modeling study using CAMx with modified volatility basis set scheme. Atmos Chem Phys. 2019;19(24):15247–70. https://doi.org/10.5194/ACP-19-15247-2019.

Kumar V, Giannoukos S, Haslett SL, Tong Y, Singh A, Bertrand A, Lee CP, Wang DS, Bhattu D, Stefenelli G, Dave JS, Puthussery JV, Qi L, Vats P, Rai P, Casotto R, Satish R, Mishra S, Pospisilova V, et al. Highly time-resolved chemical speciation and source apportionment of organic aerosol components in Delhi, India, using extractive electrospray ionization mass spectrometry. Atmos Chem Phys. 2022;22(11):7739–61. https://doi.org/10.5194/ACP-22-7739-2022.

Manisalidis I, Stavropoulou E, Stavropoulos A, Bezirtzoglou E. Environmental and health impacts of air pollution: a review. Front Public Health. 2020;8:14. https://doi.org/10.3389/fpubh.2020.00014.

Mishra S, Tripathi SN, Kanawade VP, Haslett SL, Dada L, Ciarelli G, Kumar V, Singh A, Bhattu D, Rastogi N, Daellenbach KR, Ganguly D, Gargava P, Slowik JG, Kulmala M, Mohr C, El-Haddad I, Prevot ASH. Rapid night-time nanoparticle growth in Delhi driven by biomass-burning emissions. Nat Geosci. 2023;16(3):224–30. https://doi.org/10.1038/s41561-023-01138-x.

Moschos V, Schmale J, Aas W, Becagli S, Calzolai G, Eleftheriadis K, Moffett CE, Schnelle-Kreis J, Severi M, Sharma S, Skov H, Vestenius M, Zhang W, Hakola H, Hellén H, Huang L, Jaffrezo

JL, Massling A, Nøjgaard JK, et al. Elucidating the present-day chemical composition, seasonality and source regions of climate-relevant aerosols across the Arctic land surface. Environ Res Lett. 2022;17(3):034032. https://doi.org/10.1088/1748-9326/AC444B.

Myhre G, Myhre CEL, Samset BH, Storelvmo T. Aerosols and their relation to global climate and climate sensitivity | Learn Science at Scitable. Nature Education; 2013. https://www.nature.com/scitable/knowledge/library/aerosols-and-their-relation-to-global-climate-102215345/.

Penner Lead JE, Andreae M, Annegarn H, Barrie L, Feichter J, Hegg D, Jayaraman A, Leaitch R, Murphy D, Nganga J, Pitari G, Ackerman A, Adams P, Austin P, Boers R, Boucher O, Chin M, Chuang C, Collins B, et al. Aerosols, their direct and indirect effects 5 contents. IPCC; 2018.

Rai P, Furger M, El Haddad I, Kumar V, Wang L, Singh A, Dixit K, Bhattu D, Petit JE, Ganguly D, Rastogi N, Baltensperger U, Tripathi SN, Slowik JG, Prévôt ASH. Real-time measurement and source apportionment of elements in Delhi's atmosphere. Sci Total Environ. 2020;742:140332. https://doi.org/10.1016/J.SCITOTENV.2020.140332.

Rantanen M, Karpechko AY, Lipponen A, Nordling K, Hyvärinen O, Ruosteenoja K, Vihma T, Laaksonen A. The Arctic has warmed nearly four times faster than the globe since 1979. Commun Earth Environ. 2022;3(1):1–10. https://doi.org/10.1038/s43247-022-00498-3.

Tobler A, Bhattu D, Canonaco F, Lalchandani V, Shukla A, Thamban NM, Mishra S, Srivastava AK, Bisht DS, Tiwari S, Singh S, Močnik G, Baltensperger U, Tripathi SN, Slowik JG, Prévôt ASH. Chemical characterization of PM2.5 and source apportionment of organic aerosol in New Delhi, India. Sci Total Environ. 2020;745:140924. https://doi.org/10.1016/J.SCITOTENV.2020.140924.

Tong Y, Pospisilova V, Qi L, Duan J, Gu Y, Kumar V, Rai P, Stefenelli G, Wang L, Wang Y, Zhong H, Baltensperger U, Cao J, Huang RJ, Prévôt ASH, Slowik JG. Quantification of solid fuel combustion and aqueous chemistry contributions to secondary organic aerosol during wintertime haze events in Beijing. Atmos Chem Phys. 2021;21(12):9859–86. https://doi.org/10.5194/ACP-21-9859-2021.

University of Leeds. Why are aerosols important for climate change, air quality and weather? | Aerosols and Climate | School of Earth and Environment | University of Leeds. University of Leeds; 2023. https://environment.leeds.ac.uk/atmospheric-chemistry-aerosols/doc/aerosols-climate/page/5.

Vogel A, Dällenbach K, El-Haddad I, Wendl I, Eichler A, Schwikowski M. Reconstruction of Aerosol concentration and composition from Glacier Ice Cores. Geophys Res Abstr. 2017;19:23–39.

Yan C, Shen Y, Stolzenburg D, Dada L, Qi X, Hakala S, Sundström AM, Guo Y, Lipponen A, Kokkonen TV, Kontkanen J, Cai R, Cai J, Chan T, Chen L, Chu B, Deng C, Du W, Fan X, et al. The effect of COVID-19 restrictions on atmospheric new particle formation in Beijing. Atmos Chem Phys. 2022;22(18):12207–20. https://doi.org/10.5194/ACP-22-12207-2022.

Extreme Heat in South Asia: Meteorology, Prediction, and Impacts

4

Joy Merwin Monteiro and Hardik Shah

Abstract

The importance of heat stress in public and occupational health in South Asia is now generally recognized. The region experiences a monsoonal climate, resulting in a stark contrast between dry and wet seasons. This strong seasonality leads to a heat exposure pattern in South Asia that differs from regions without a monsoonal climate. To identify and evaluate adaptation measures for extreme heat, it is essential to understand both the meteorology of extreme heat and its impacts on the human body. This chapter provides an overview of both aspects—meteorology and impacts—with the aim of delineating areas of significant overlap. It then outlines challenges that must be addressed to improve systems-level forecasting and adaptation to extreme heat in a changing climate in South Asia.

Keywords

Extreme heat · Meteorology · South Asia · Prediction · Heat-Health System

4.1 Introduction

As the climate warms, the likelihood of exceptionally warm periods or heat waves is projected to increase (Fischer et al. 2021). This makes adapting to extreme heat a high priority across the world, particularly in South Asia, which is home to a large population vulnerable to the impacts of extreme heat (Romanello et al. 2021). In contrast to the more dramatic visual impacts of other natural hazards such as

J. M. Monteiro (✉) · H. Shah
Department of Earth & Climate Science, Indian Institute of Science Education Research Pune, Pune, India
e-mail: joy@iiserpune.ac.in

H. Brand et al. (eds.), *Climate Change and Public Health*,
https://doi.org/10.1007/978-981-95-0895-2_4

droughts, floods, and cyclones, the perception of extreme heat and its impacts can be invisible to a large extent (Moore et al. 2019), and not just because those primarily impacted by extreme heat are the poor and marginalized (Azhar et al. 2017).

In this chapter, we aim to present the biophysical underpinnings of the relationship between humans and their thermal environment. A clear understanding of this relationship is essential to understanding the strengths and limitations of different ways and means to predict, adapt, and build resilience to extreme heat. A conceptual understanding of the exchange of heat between the human body and the environment and the environmental factors that control this exchange was present for at least a hundred years in South Asia—Normand (1920) tried to understand why hot, dry winds of the Middle East could be fatal and went on to develop a clear description of the ambient conditions under which the human body will be unable to maintain a stable core temperature. However, translating this conceptual understanding into an operational framework that protects individuals and communities from the effects of extreme heat remains a challenge to this day. Part of the challenge lies in the chaotic nature of weather, which ensures that weather forecasts (especially of extreme events) are always uncertain to some degree, and part of it lies in the wide variety of factors, most of them not well measured, that shape the response to extreme heat. In this chapter, we focus on two necessary aspects to enable this translation—the meteorology of extreme heat and its impact on human physiology—and review the progress that has been made in South Asia and the challenges that remain.

4.2 Human-Environment Energy Balance

To frame the subsequent discussion, it is necessary to build a conceptual picture of why and how ambient heat affects human physiology. Figure 4.1 is a schematic that outlines the basic thermodynamic picture of the human-environment heat exchange. The metabolic energy generation rate M is partially used to perform work W_k which includes autonomic bodily functions and other activities such as walking, exercise, and physical labor, and the remaining is converted to heat, which must be removed from the body to prevent accumulation of heat.[1] The efficiency by which M is converted to W_k depends on the kind of physical activity performed and individual physical fitness (Cramer and Jay 2016).

There are four pathways for the body to lose this excess heat, all of which are familiar from elementary thermodynamics: conduction, convection, evaporation, and radiation. As we will discuss subsequently, conduction, convection, and radiation are also pathways by which the body is supplied heat from the environment, whereas evaporation can only result in the loss of heat from the body. Since air is a poor conductor of heat, the loss of heat due to conduction can be ignored for common scenarios.

[1] We focus only on situations where heat accumulation is an issue, and not heat loss which can occur in cold conditions.

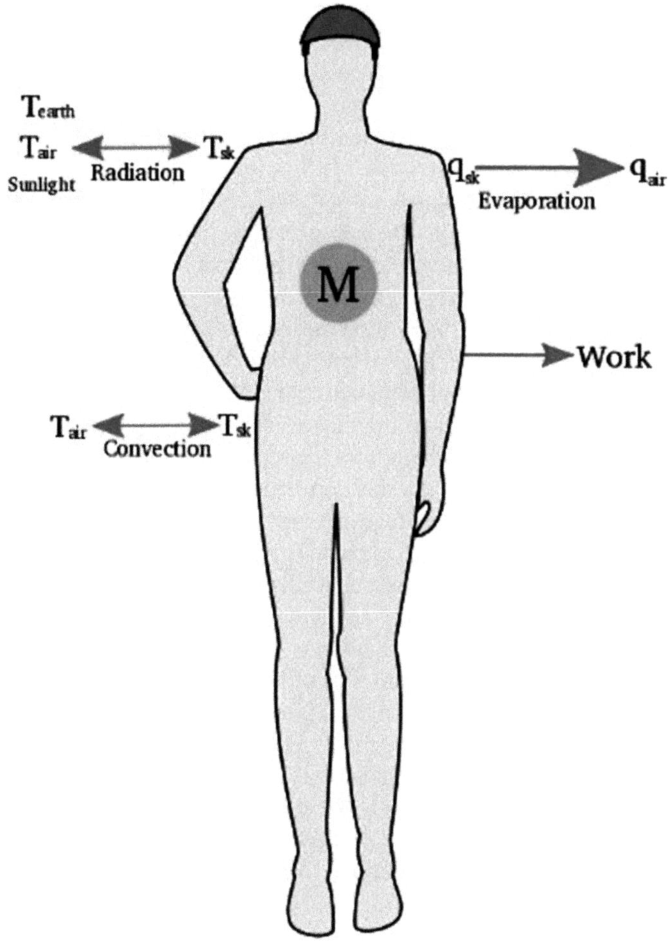

Fig. 4.1 The energy balance of the human body for its environment. *M* represents the metabolic energy generation rate. Orange arrows represent energy pathways that can both reject and add heat to the body. Green arrows represent energy pathways that primarily reject heat from the body. (Figure adapted from https://publicdomainvectors.org/en/free-clipart/Human-body-silhouette-vectpr/8390.html)

Convective heat loss depends on the difference between the ambient temperature T_{air} and the skin temperature T_{sk}, the magnitude of winds, and the resistance to loss of heat (Caballero 2014). This resistance can be attributed to the properties of clothing worn by an individual—protective clothing, in particular, can dramatically reduce loss of heat due to convection. Furthermore, if T_{air} is greater than T_{sk}, convection supplies heat to the body, and the rate of this supply of heat is accentuated by increasing wind speed. This is the reason why very hot and windy days in deserts can be hazardous (Normand 1920). The difference between T_{sk} and the core temperature T_c determines the rate at which heat is transferred from the core to the skin where it is rejected by the environment. Again, if T_{sk} is greater than T_c, heat flows to the core. Therefore, from the point of view of heat transfer, the skin temperature T_{sk} is an important factor since it determines the rate of heat transfer both to the environment and from the core. Convective heat transfer also happens from the respiratory tract, but its magnitude is small compared to the convective heat transfer from the skin (Cramer and Jay 2016).

As we know well from elementary physics, any object that has a temperature emits electromagnetic radiation. The rate of energy emitted by radiation is proportional to the fourth power of the temperature of the object (Caballero 2014). In the context of the human heat budget, radiation from the sun and from the earth's surface and the atmosphere are incident on the skin and increase the supply of heat to the body. The skin itself loses heat by radiation which depends on T_{sk}. On regular, cloud-free days, solar radiation dominates the radiative budget. The supply of heat due to solar radiation depends on the time of the day and the presence/absence of shade. During heat waves, radiation from the earth's surface and the atmosphere can also become significant and add to the convective heat load described in the previous paragraph. In certain occupational settings, the presence of furnaces, boilers, or kilns can lead to additional radiative heat load. Radiative heat load can depend quite dramatically on the built environment as well—for instance, not only is there a heat island effect, which is an increase in ambient temperatures T_{air} in concretized urban regions, but heat supplied due to radiation is also higher, and more sustained since concrete structures absorb and retain a lot of heat which is radiated for many hours after the peak in solar radiation (Tan et al. 2014).

Evaporation is the only pathway that is unidirectional—the body can only lose heat due to evaporation and cannot gain heat. Like convective heat transfer, evaporative heat loss depends on the ambient-specific humidity q_a, the moisture availability at the skin (measured as "skin wettedness," Cramer and Jay 2016), wind speeds, and resistance to evaporation due to clothing. There is a small contribution of evaporation of water from the respiratory tract that we will not consider here. We note that specific humidity q_a—the mass of water vapor in one kilogram of air—is important in determining the rate of evaporative heat transfer and rarely used humidity variables such as relative humidity or dew-point temperature. These latter measures of humidity implicitly depend on temperature as well and are not useful in determining the impact of humidity on the heat budget of the human body (Baldwin et al. 2023). Therefore, it is important to use the correct measure of humidity in epidemiological studies. Since the latent heat of vaporization of water is very high, a small amount of sweat evaporating can result in substantial removal of heat from the body. Therefore, the evaporation of sweat is a very efficient way of losing heat in the human body. Moisture supply and availability on the skin can be reduced when the person is dehydrated, aged, or has conditions that reduce the ability to sweat such as diabetes (Ebi et al. 2021). In hot, dry conditions, this factor usually determines the risk to health since the amount of evaporative heat loss is determined by the ability to produce sweat. On the other hand, increasing values of q_{air} (which represent very humid conditions) reduce evaporative efficiency by reducing the amount of sweat that evaporates. Therefore, even if a person is healthy and fully hydrated and can produce an adequate amount of sweat, evaporative heat loss may not be sufficient to remove heat from the body since the sweat does not evaporate and remove heat. Thus, in hot, humid conditions, evaporative efficiency may be determined by ambient conditions rather than physiological limitations.

Given the unidirectionality and efficiency of sweat evaporation in heat removal, it is worthwhile to club together all other terms of the heat budget into one value and see if it can be balanced by evaporative heat loss. This required rate of evaporative heat loss E_R needed to balance heat supply/production in the body can be then compared with the theoretical maximum evaporative heat loss E_M, which is determined by ambient air temperature, humidity, and air pressure. E_M decreases when the ambient humidity increases since the atmosphere supports lesser and lesser evaporative rates. If E_R is lesser than E_M, then the person is said to experience "compensable" heat stress and is said to experience "uncompensable" heat stress otherwise (Cramer and Jay 2016; Guzman-Echavarria et al. 2022). To state in another way, uncompensable heat stress is the condition where there is a net accumulation of heat in the human body, and this condition is determined by both the ambient environmental conditions as well as personal attributes such as rate of physical activity, clothing, and morbidities. When the ratio E_R/E_M exceeds 0.5, which can happen when the ambient conditions are very humid, the efficiency of sweating in heat removal decreases and such conditions cause high heat stress (Cramer and Jay 2016; Guzman-Echavarria et al. 2022). Since E_R depends on both ambient conditions and the rate of physical activity and morbidities could exist which inhibit sweat production, the same ambient conditions can be compensable for one person and uncompensable for another, a fact that was recognized very early by Normand (1920).

The concept of uncompensable heat stress can also be used to clarify the role of wet-bulb temperatures which have received considerable scholarly and media attention over the past decade. Wet-bulb temperature is the temperature that is attained by a thermometer whose bulb is covered by a cloth that is continuously moistened. Since part of the heat supplied to the bulb is lost due to evaporation, the temperature measured by the wet-bulb thermometer is less than or equal to the temperature measured by a regular (or dry-bulb) thermometer. In a hypothetical highly humid environment where loss of heat due to evaporation is impossible, both dry- and wet-bulb temperatures are the same. In contrast, in a very dry environment, heat can be lost due to evaporation very efficiently and the wet-bulb temperature is much lower than the dry-bulb temperature. This reduction in the efficiency of sweating and the resulting uncompensable heat stress has led to suggestions that sustained wet-bulb temperatures of 35 °C represent the limit of adaptability of humans to heat stress (Sherwood and Huber 2010) and climate change projections suggest that wet-bulb temperatures will approach this limit in South Asia in this century (Im et al. 2017; Pal and Eltahir 2016; Raymond et al. 2020).

Thus, high values of wet-bulb temperature indicate both hot and humid (or equivalently, high heat stress) conditions. However, it does not mean that hot and dry conditions characterized by lower wet-bulb temperatures are less hazardous to human health. In very hot conditions, the heat supplied to the body due to convection and radiation may be so high as to result in uncompensable heat stress even if the efficiency of sweat evaporation is high. This is likely the reason why epidemiological studies find significant associations of mortality with dry-bulb temperatures in South Asia and other parts of the world (Baldwin et al. 2023; Dimitrova et al. 2021), and most deadly heat waves in South Asia are associated with lesser than normal wet-bulb temperatures (Justine et al. 2023). Thus, instead of extreme dry- and wet-bulb temperatures, focusing on compensable and uncompensable heat stress provides a more physiologically relevant way of thinking about the impact of heat on health.

Given this simplified model of the energy balance between the human body and the environment, we next look at the climatology of the basic environmental factors—temperature, humidity, winds, and radiation—that are involved in this balance.

4.3 Meteorology of South Asia

South Asia experiences a monsoonal climate, which means that there is a stark difference between dry and wet seasons. Monsoonal regions are different from other tropical regions, which do not have a pronounced seasonal cycle in temperature or precipitation, or from the extra tropics which have a strong seasonal cycle of temperature but a relatively muted seasonal cycle in precipitation. As we will see, this strong seasonality leads to a heat exposure pattern in South Asia which is different from other parts of the world that are not monsoonal.

4.4 Radiation

The seasonal cycle of the net solar radiation—the difference between incident and reflected solar radiation at a location—is presented in Fig. 4.2. While the maximum radiation in the northern hemisphere is expected during the solstice in June, the maximum over South Asia is reached in May and dramatically reduces in June. This is due to the march of the monsoon, which results in increasing cloudiness after the onset in June as seen in Fig. 4.2. As the monsoon covers most of South Asia by July, solar radiation drops to values that are much smaller than in April, which is close to the equinox. However, the Indus Valley region in Pakistan continues to experience high solar radiation in June as well, since the monsoon onset over this region is in July (Janowiak and Xie 2003). While we have not shown the seasonal changes in thermal radiation, it closely follows that of solar radiation since it depends on the temperature of the land surface and the near-surface air, both of which are heated primarily by the sun.

Therefore, while most of the northern hemisphere experiences the highest annual radiation during the traditionally defined summer season of June–August, South Asia experiences the highest radiation just before June, and it is in this season (April–June) when most heat- waves in this region occur. Using this same line of reasoning, delays in monsoon onset or interruptions in the monsoon—which occur

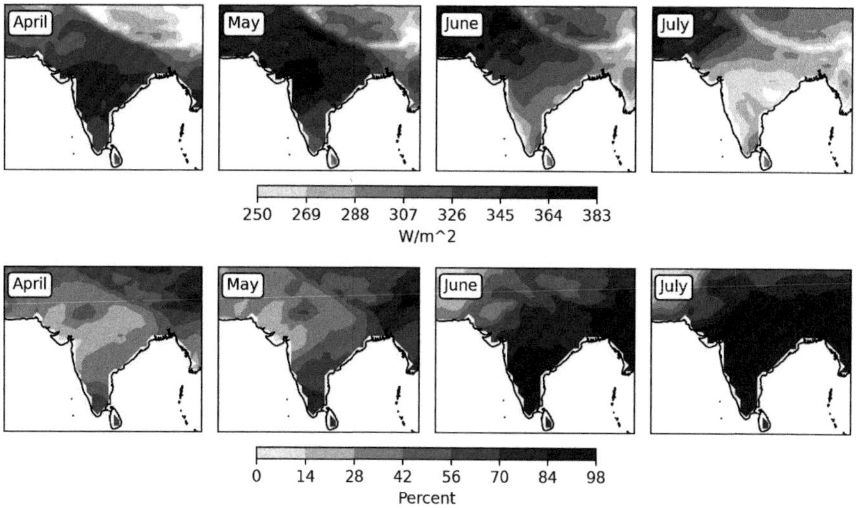

Fig. 4.2 The monthly mean net solar radiation (top row) and cloud area fraction (bottom row) over South Asia. (Source: CERES dataset (Wielicki et al. 1996))

for instance during an El-Niño year—reduce cloudiness during this season and can lead to high heat exposure conditions even during June and July.

4.5 Temperature, Humidity, and Winds

Since solar radiation is the main source of energy for heating the earth's surface and the atmosphere above it, the seasonal cycle of the near-surface air temperature is similar to that of radiation, with maximum temperatures being reached earlier in peninsular India as compared to north India and Pakistan, as seen in Fig. 4.3. The highest mean temperatures are reached in the interior regions—central and northwest parts of South Asia, away from the moderating influence of the ocean and the cooling effect of increasing elevation. With the arrival of the monsoon, near-surface temperatures reduce not only due to clouds cutting off solar radiation but also the cooling effect of raindrops evaporating as they fall to the ground. Thus, there is a dramatic reduction in mean temperatures in regions where the monsoon is active.

Monsoonal regions exhibit a dramatic change in wind patterns over the year as seen in Fig. 4.4. While the near-surface winds are expected to be blowing west to east in concordance with the trade winds in the rest of the tropics, they reverse direction nearly a month before the monsoon in May, and the direction of winds is almost completely reversed by June, especially in the Ganga plains. Not only do the winds reverse direction, but they also increase in magnitude. Therefore, not only is there a decrease in temperatures but also an increase in wind speeds which mitigates heat exposure.

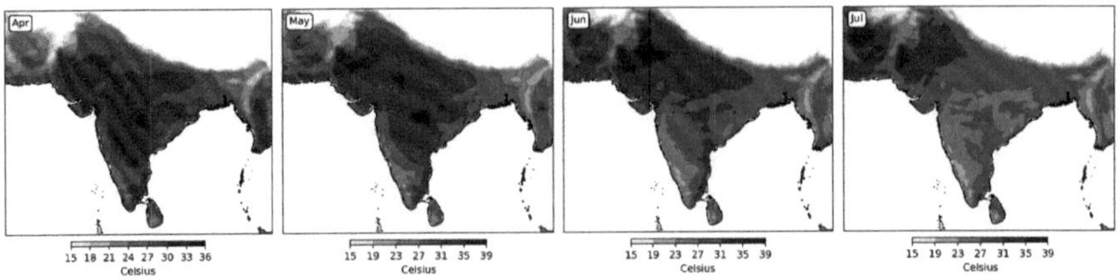

Fig. 4.3 The monthly mean 2-metre air temperature. (Source: ERA-5 dataset (Hersbach et al. 2020))

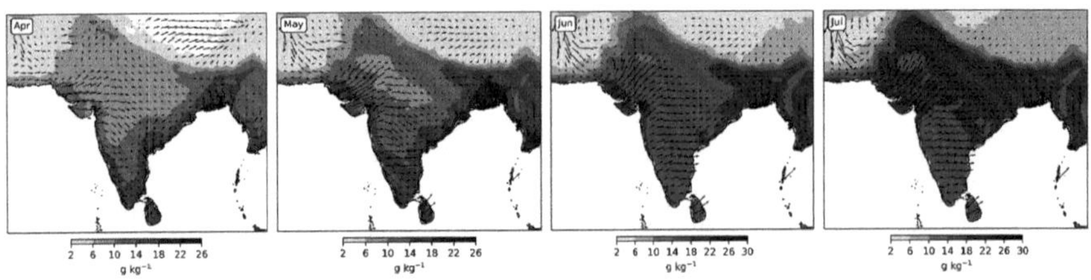

Fig. 4.4 The monthly mean 2-metre specific humidity (shaded colors) and winds (arrows). (Source: ERA-5 dataset)

Humidity exhibits a more complex distribution over the season. While the coastal regions are more humid as expected, Fig. 4.4. shows the existence of a few tongues of high humidity that intrude into South Asia in Andhra Pradesh and the Indus and Ganga plains. The humid tongue over the Ganga plains becomes longer as we approach July. These tongues of humidity are likely the result of the long-range transport of water vapor from the ocean by the monsoon winds. The extent of these tongues is determined by the extent to which monsoonal winds can intrude into the South Asian landmass. Since wet-bulb temperatures are very sensitive to changes in humidity (Monteiro and Caballero 2019), the distribution of high wet-bulb temperatures across South Asia mirrors the distribution of humidity presented in Fig. 4.4. (Im et al. 2017).

In summary, even before the onset of the monsoon, there is a rise in humidity in Andhra Pradesh and the Indus-Ganga plains; along with the prevailing high levels of solar radiation and high temperatures, these regions experience conditions that reduce both efficiency and magnitude of energy loss from the human body and can thus be termed as regions of high heat stress. On the other hand, central and northwest parts of South Asia experience extremely hot and dry conditions; while these regions also experience high heat stress, this is primarily due to an increase in energy input to the body rather than a reduction in the efficiency of energy loss.

4.5.1 Heat Waves

A heat wave is broadly defined as an extended period of exceptional near-surface temperatures that usually last for a week or less. Heat waves have been traditionally defined using dry-bulb temperatures, though these may not directly translate to health impacts. As discussed in the previous section, the pre-monsoon season in South Asia is characterized by climatologically high heat stress, and therefore, heat wave conditions in South Asia can be thought of as the proverbial last straw on the camel's back.

4.5.2 Why Do Heat Waves Occur?

Every location on the planet is characterized by a temperature that depends on the amount of heat supplied to it and removed from it. The heat is supplied primarily by solar radiation which depends on the day of the year and is removed by thermal radiation and exchange of energy with the atmosphere above it in the form of convection and evaporation. This picture of the energy balance of a geographical location corresponds quite nicely with the picture we described for the human body earlier—this shows how powerful and generalizable energy balance techniques can be. Figure 4.5 outlines the basic mechanisms that lead to heat waves and shows how heat waves are the result of excess input or reduced removal of energy at a location—in some sense a condition of uncompensable heat stress for landscapes rather than humans.

Understanding how heat waves change as the climate warms then becomes a question of understanding how the mechanisms that disrupt a location's energy

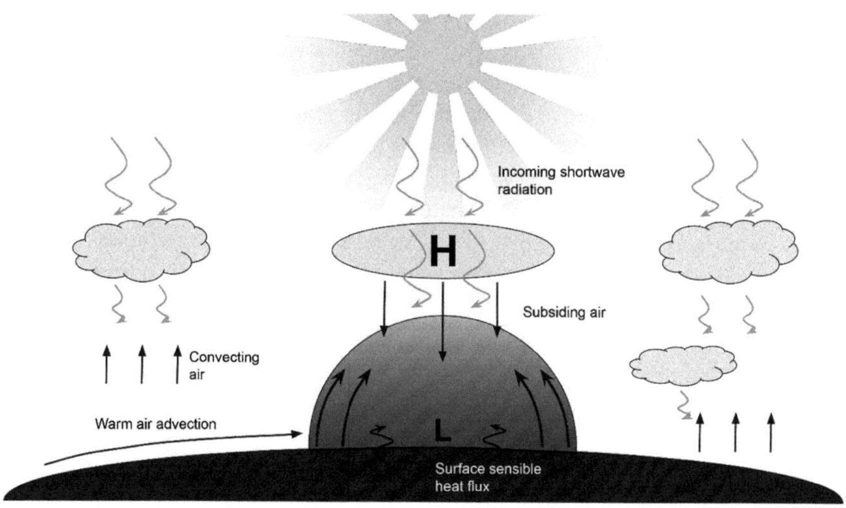

Fig. 4.5 Heat waves are a direct result of the accumulation of heat near the surface. One of the main atmospheric features accompanying such conditions is a high-pressure system aloft, directly over or in the neighborhood of, the region experiencing the heat wave. Such a system restricts the formation of clouds and promotes clear skies, which allow a lot more incoming shortwave radiation to pass. Accordingly, the surface gets heated, and in turn, the near-surface air. Additionally, the upper-level air tends to subside, restricting the near-surface hot air from rising and dissipating the heat. If the atmospheric conditions at night do not allow this air mass to mix with the rest of the atmosphere, the near-surface heat accumulation from 1 day can carry over to the next, considerably amplifying the heat wave intensity

balance will change. Model studies indicate that a warmer climate primarily caused by greenhouse gas-driven warming of the troposphere can enhance the likelihood and intensity of heat waves. The increase in likelihood can be attributed solely to an increase in background temperatures for some regions (Cowan et al. 2014), while for some others, the associated circulation changes can cause an increase in the frequency of heat wave-causing conditions (Diffenbaugh and Ashfaq 2010). The rate at which global warming occurs has been shown to influence how often extreme heat waves will occur over the coming century (Fischer et al. 2021). Specific humidity in the atmosphere is also likely to increase with increased temperatures, which will enhance heat stress during heat waves (Lutsko 2021).

4.5.3 Where Do Heatwaves Occur?

South Asian regions are affected differently by different large-scale modes of climate variability, giving rise to a variety of heat wave characteristics that impact heat stress differently. To look at the distribution of extreme heat in South Asia, we use the Excess Heat Factor Index (EHF). EHF has been shown to correlate well with human health impacts, especially in Australia where it has been developed (Nairn et al. 2018). EHF captures unusually high heat arising from a daytime temperature that is not sufficiently discharged overnight due to unusually high overnight temperatures, using the mean of maximum and minimum daily temperature, called daily mean temperature. The index is designed to capture the extent of positive deviation of daily mean temperature from both historical long-term (March–June) and short-term (30-day) temperatures at a given location, on any given day. Due to its design which is intended to capture the warmest days in the entire year, high EHF values in South Asia mostly occur during the hottest months of the pre-monsoon period, with very few, exceptionally hot, periods identified in March and April. Since most days are close to the climatological mean, extreme values of EHF are observed at regions and times that experience extremely warm conditions.

The most intense heat waves are found to occur in east Pakistan, northwest India, northeast Indian regions around northern West Bengal, and the southeast Indian coastal state of Andhra Pradesh (Fig. 4.6). Both the mean and standard deviation of EHF are higher in these three regions than in other parts of South Asia, suggesting a tendency to experience extreme heat waves. The largest daily mean temperatures are observed in northwest India and east Pakistan, where intense heat waves add to the hot seasonal conditions, leading to extreme temperatures. These regions also experience the highest frequency of heat waves (Fig. 4.7, right plot), and therefore, extreme heat affects these regions most often as com pared to the rest of South Asia.

The mean and standard deviation of heat wave durations (Fig. 4.7, left and middle plots) imply that heat wave duration is very consistently below 10 days in the northwest and northeast Indian regions and can reach up to more than 25 days in the southeast Indian hotspot. Some parts of Karnataka can experience heat waves of a duration of more than 35 days, but their intensity is so low that they are not comparable with the intensities found in the three hotspots mentioned previously.

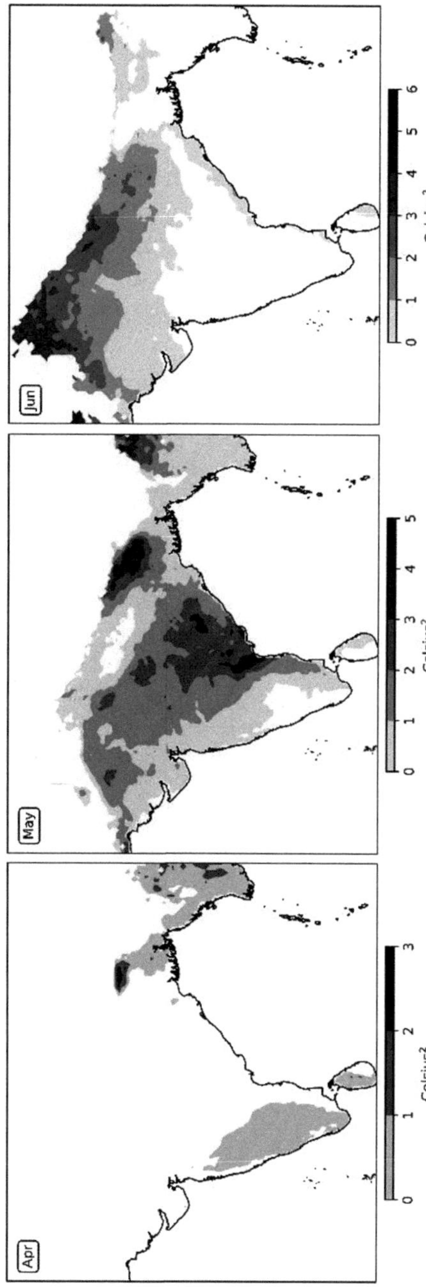

Fig. 4.6 The monthly 95th percentile of EHF per region. (Source: ERA-5 dataset (Hersbach et al. 2020))

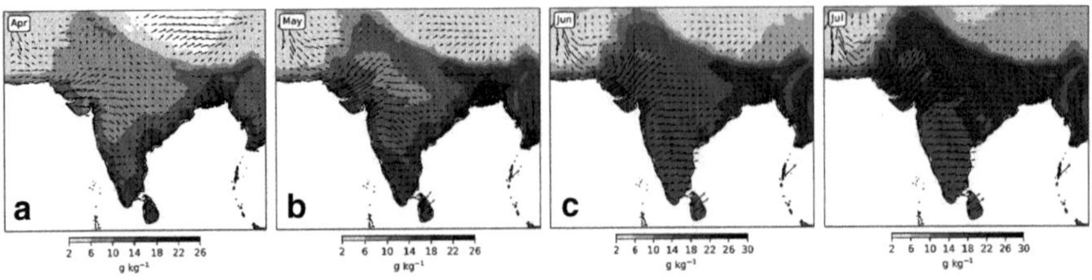

Fig. 4.7 EHF-based heat wave characteristics (**a**) mean duration, (**b**) standard deviation of duration, and (**c**) heat wave counts, in the period 1980–2022. The boxes mark the locations of the three hotspots described in the text

4.6 Trends and Projections Over South Asia

While multiple studies have studied heat wave trends in observational data and climate projections, quantitative comparison between these studies is hindered due to the lack of a standard method to define heat waves and heat wave characteristics such as intensity, duration, and frequency. While quantitative comparison is difficult, one of the advantages of using different metrics is that estimated changes can be considered more robust when more than one study arrives at the same result. In the case of South Asia (in fact, globally), it is indeed the case that there is an unambiguous trend in heat wave frequency and duration across multiple studies described below.

Heat waves over South Asia have been traditionally identified using dry-bulb temperature alone. India has a long and reliable instrumental record of air temperature which allows for estimating historical trends of heat waves across the subcontinent. Analysis of the long Indian temperature record suggests that two of the three heat wave hotspots in South Asia—northwest and southeast—show increasing trends in frequency and maximum duration of individual heat waves and seasonal duration of heat wave days (Dash and Mamgain 2011; Rohini et al. 2016). A similar analysis from Pakistan also shows increasing trends in heat wave indices (Khan et al. 2018). Even in global analyses of heat wave characteristics, South Asian trends in total heat wave days and duration of the longest heat wave are among the highest in the world (Perkins-Kirkpatrick and Lewis 2020), while following the global trend of having no appreciable trends in heat wave intensity.

Climate projections using CMIP5[2] data suggest that while heat wave intensity and duration over South Asia are projected to increase over this century, these trends become small in scenarios where the global mean temperature is stabilized at 1.5 °C by the end of the century. RCP8.5 (or "business as usual") scenarios suggest a dramatic increase in all heat wave characteristics, but the utility of this scenario in terms of adaptation planning is being debated (Hausfather and Peters 2020; Schwalm

[2]The Coupled Model Intercomparison Project 5 is a global initiative that produces climate model simulations for well-defined future scenarios with the aim of providing robust climate projections to inform the science and policy of climate change (Taylor et al. 2012).

et al. 2020a, b). Some studies estimate changes in mortality with future trends in heat waves (Mazdiyasni et al. 2017; Murari et al. 2015), but such studies use regression or correlation-based methods which are considered inappropriate for estimating relationships between temperature and mortality (Dimitrova et al. 2021).

In summary, there seems to be a consensus that heat wave frequency and duration have been increasing and will continue to increase over the coming century. However, the potential health impacts are poorly understood due to a lack of quality health data to understand current risks as well as methodological issues in extrapolating current risks to future climates.

4.7 Assessment of Human Health Impacts

4.7.1 Measurement

Reliable meteorological forecasts and climate projections are the first steps in any heat-health decision-making framework. The second step is assessing whether the predicted atmospheric conditions are hazardous, and to what extent. This assessment can be approached from two different methodologies: one, using historical data of health outcomes (such as all-cause mortality or hospital admissions) and atmospheric conditions along with statistical methods to determine thresholds of meteorological variables such as temperature and humidity at which adverse outcomes begin to rise rapidly. In India and globally, air temperature has been successfully used as the predictive variable even though as we have seen in a previous section, it is only one component of the human energy balance (Baldwin et al. 2023; Dimitrova et al. 2021). Two, using physiological experiments or models to determine ranges of meteorological variables that correspond to high levels of physiological strain or conversely atmospheric conditions corresponding to high heat stress. The former approach has been the default practice, including in South Asia, and is described elsewhere in this book. In this chapter, we present the latter approach as a more suitable alternative given the diversity of climatological and epidemiological conditions prevalent in India.

Physiological heat stress is usually measured by indices that combine the multiple atmospheric variables above into a single number. These indices are called thermal stress indicators (TSIs), and a wide variety of TSIs have been developed (Ioannou et al. 2022). TSIs can be used to estimate instantaneous heat stress and therefore can allow for intra-day decision making which is useful in occupational and sports settings (Havenith and Fiala 2015). Popularly used TSIs include Wet-bulb Globe Temperature (WGBT), Universal Thermal Climate Indicator (UTCI), Apparent Temperature, and Humidex, though it is very important to understand the validity of these TSIs for any particular application—indoors versus outdoors, low versus high thermal stress, for example—before using them (Buzan et al. 2015; Havenith and Fiala 2015; Ioannou et al. 2022; Kong and Huber 2022; Simpson et al. 2023; Vanos et al. 2020). The fact that TSIs can be computed using mathematical equations makes them attractive for modelling, scenario building, and future health risk projections since epidemiological approaches cannot be easily applied in such applications. As climate data become more easily available at high space and time resolutions, TSIs will likely be used for population- and regional-scale risk assessments as well (Glass et al. 2015; Guzman-Echavarria et al. 2022; Justine et al. 2023). Thresholds established using epidemiological analyses are usually at daily timescales and therefore cannot be used to estimate instantaneous heat stress. On the other hand, epidemiological thresholds are based on actual health outcomes and therefore indirectly consider many other populations' risk factors (demographic, socio-economic, among others), which are usually not reflected in TSIs.

4.8 Conclusion

Managing and adapting to extreme heat is a systems problem. An effective heat-health system focuses on

Better prediction of extreme heat and related climatic variables of relevance to health (humidity, radiation, and wind speed)

Better translation of these climatic variables into potential health risks

Better translation of health risk predictions into interventions on the ground

Improving the ability of vulnerable groups to participate in such interventions

In the context of public health, such a system represents a primary prevention measure, unlike secondary prevention measures such as surveillance and response more popular in the management of infectious diseases (Ebi 2005). While many public health programs are programmatic, such an approach is unlikely to be very effective in dealing with environmental stressors where an understanding of the vulnerability of particular populations and groups is important (Sen et al. 2017), making local action and community participation an integral part of the solution (Devine-Wright et al. 2022). In the same vein, understanding the influence of the built environment in determining vulnerability and response to extreme heat is essential (Kotharkar and Ghosh 2021).

Predictions of extreme heat have been continuously improving as computing infrastructure has improved in India over the past 15 years (Mandal et al. 2019; Rao et al. 2019; Rohini et al. 2022). The availability of high-quality health data has allowed the translation of such predictions into health risks in some locations (Azhar et al. 2014; Ingole et al. 2012; Knowlton et al. 2014; Nori-Sarma et al. 2019), leading to effective heat action plans. However, such approaches do not seem to be scalable, and most heat action plans make simplistic assumptions about the relationship between heat and health impacts. This issue is further compounded by the lack of vulnerability assessments (Pillai and Dalal 2023). Use of TSIs has been a challenge even in household/occupational settings in India and occupational exposure compounded by environmental exposure and its impacts on health are virtually unknown in South Asia (Krishnamurthy et al. 2017; Mukhopadhyay and Weitz 2022; Venugopal et al. 2016, 2021a, b; Weitz et al. 2022). Furthermore, the use of TSIs at the population level is still in its infancy due to methodological issues regarding the choice of TSIs (Kong and Huber 2022; Lu and Romps 2022; Simpson et al. 2023) and their implications for the regional level exposures (Glass et al. 2015; Guzman-Echavarria et al. 2022; Justine et al. 2023). Translation of predictions to interventions is documented in heat action plans, but the inherent uncertainty in weather forecasts makes roll-out of certain interventions (such as stockpiling of medicines and recruiting additional health personnel) a tricky affair. Decision-making under uncertainty given the cost (both material and social) of false positives or negatives is not very well understood (Jewson et al. 2021).

Finding ways and means to adapt to extreme heat will likely remain a challenge in the coming decade not only in South Asia but worldwide. Promising avenues for future research and practice are available, and this chapter aims to describe some of the challenges and avenues for future work. The bibliography accompanying this chapter should provide a good starting point for both understanding heat waves and their impacts in South Asia. The interdisciplinary nature of understanding and response to environmental stressors makes the study of heat and health as well as the policy response an exciting and important mission for the coming decades.

References

Azhar GS, Mavalankar D, Nori-Sarma A, Rajiva A, Dutta P, Jaiswal A, Sheffield P, Knowlton K, Hess JJ. Heat-related mortality in India: excess all-cause mortality associated with the 2010 Ahmedabad Heat Wave. PLoS One. 2014;9(3):e91831. https://doi.org/10.1371/journal.pone.0091831.

Azhar G, Saha S, Ganguly P, Mavalankar D, Madrigano J. Heat wave vulnerability mapping for India. Int J Environ Res Public Health. 2017;14(4):357. https://doi.org/10.3390/ijerph14040357.

Baldwin JW, Benmarhnia T, Ebi KL, Jay O, Lutsko NJ, Vanos JK. Humidity's role in heat-related health outcomes: a heated debate. Environ Health Perspect. 2023;131(5):055001. https://doi.org/10.1289/EHP11807.

Buzan JR, Oleson K, Huber M. Implementation and comparison of a suite of heat stress metrics within the Community Land Model version 4.5. Geosci Model Dev. 2015;8(2):151–70. https://doi.org/10.5194/gmd-8-151-2015.

Caballero R. Physics of the atmosphere. IOP PUBL Limited; 2014.

Cowan T, Purich A, Perkins S, Pezza A, Boschat G, Sadler K. More frequent, longer, and hotter heat waves for Australia in the twenty-first century. J Clim. 2014;27(15):5851–71. https://doi.org/10.1175/JCLI-D-14-00092.1.

Cramer MN, Jay O. Biophysical aspects of human thermoregulation during heat stress. Autonom Neurosci Basic Clin. 2016;196:3–13. https://doi.org/10.1016/j.autneu.2016.03.001.

Dash SK, Mamgain A. Changes in the frequency of different categories of temperature extremes in India. J Appl Meteorol Climatol. 2011;50(9):1842–58. https://doi.org/10.1175/2011JAMC2687.1.

Devine-Wright P, Whitmarsh L, Gatersleben B, O'Neill S, Hartley S, Burningham K, Sovacool B, Barr S, Anable J. Placing people at the heart of climate action. PLOS Clim. 2022;1(5):e0000035. https://doi.org/10.1371/journal.pclm.0000035.

Diffenbaugh NS, Ashfaq M. Intensification of hot extremes in the United States. Geophys Res Lett. 2010;37(15):10.1029/2010GL043888.

Dimitrova A, Ingole V, Basagaña X, Ranzani O, Milà C, Ballester J, Tonne C. Association between ambient temperature and heat waves with mortality in South Asia: systematic review and meta-analysis. Environ Int. 2021;146:106170. https://doi.org/10.1016/j.envint.2020.106170.

Ebi KL. Improving public health responses to extreme weather events. In: Kirch W, Bertollini R, Menne B, editors. Extreme weather events and public health responses. Springer; 2005. p. 47–56. https://doi.org/10.1007/3-540-28862-7.

Ebi KL, Capon A, Berry P, Broderick C, de Dear R, Havenith G, Honda Y, Kovats RS, Ma W, Malik A, Morris NB, Nybo L, Seneviratne SI, Vanos J, Jay O. Hot weather and heat extremes: health risks. Lancet. 2021;398(10301):698–708. https://doi.org/10.1016/S0140-6736(21)01208-3.

Fischer EM, Sippel S, Knutti R. Increasing probability of record-shattering climate extremes. Nat Clim Chang. 2021;11(8):689–95. https://doi.org/10.1038/s41558-021-01092-9.

Glass K, Tait PW, Hanna EG, Dear K. Estimating risks of heat strain by age and sex: a population-level simulation model. Int J Environ Res Public Health. 2015;12(5):5241–5255. https://doi.org/10.3390/ijerph120505241.

Guzman-Echavarria G, Middel A, Vanos J. Beyond heat exposure — new methods to quantify and link personal heat exposure, stress, and strain in diverse populations and climates: the journal temperature toolbox. Temperature. 2022;0(0):1–21. https://doi.org/10.1080/23328940.2022.2149024.

Hausfather Z, Peters GP. Emissions—the 'business as usual' story is misleading. Nature. 2020;577(7792):618–20. https://doi.org/10.1038/d41586-020-00177-3.

Havenith G, Fiala D. Thermal indices and thermophysiological modeling for heat stress [_eprint: https://onlinelibrary.wiley.com/doi/pdf/10.1002/cphy.c140051]. In: Comprehensive physiology. Wiley; 2015. p. 255–302. https://doi.org/10.1002/cphy.c140051.

Hersbach H, Bell B, Berrisford P, Hirahara S, Horányi A, Muñoz-Sabater J, Nicolas J, Peubey C, Radu R, Schepers D, Simmons A, Soci C, Abdalla S, Abellan X, Balsamo G, Bechtold P, Biavati G, Bidlot J, Bonavita M, et al. The ERA5 global reanalysis. Q J R Meteorol Soc. 2020;146(730):1999–2049. https://doi.org/10.1002/qj.3803.

Im E-S, Pal JS, Eltahir EAB. Deadly heat waves projected in the densely populated agricultural regions of South Asia. Sci Adv. 2017;3(8):e1603322. https://doi.org/10.1126/sciadv.1603322.

Ingole V, Juvekar S, Muralidharan V, Sambhudas S, Rocklöv J. The short-term association of temperature and rainfall with mortality in Vadu Health and Demographic Surveillance System: a population level time series analysis. Glob Health Action. 2012;5(1):19118. https://doi.org/10.3402/gha.v5i0.19118.

Ioannou LG, Mantzios K, Tsoutsoubi L, Notley SR, Dinas PC, Brearley M, Epstein Y, Havenith G, Sawka MN, Bröde P, Mekjavic IB, Kenny GP, Bernard TE, Nybo L, Flouris AD. Indicators to assess physiological heat strain—Part 1: systematic review. Temperature. 2022;9(3):227–62. https://doi.org/10.1080/23328940.2022.2037376.

Janowiak JE, Xie P. A global-scale examination of monsoon-related precipitation. J Clim. 2003;16(24):4121–33. https://doi.org/10.1175/1520-0442(2003)016<4121:AGEOMP>2.0.CO;2.

Jewson S, Scher S, Messori G. Decide now or wait for the next forecast? testing a decision framework using real forecasts and observations. Mon Weather Rev. 2021;149(6):1637–50. https://doi.org/10.1175/MWR-D-20-0392.1.

Justine J, Monteiro JM, Shah H, Rao N. The diurnal variation of wet bulb temperatures and exceedance of physiological thresholds relevant to human health in South Asia. Commun Earth Environ. 2023;4(1):1–11. https://doi.org/10.1038/s43247-023-00897-0.

Khan N, Shahid S, Ismail T, Ahmed K, Nawaz N. Trends in heat wave related indices in Pakistan. Stoch Env Res Risk A. 2018; https://doi.org/10.1007/s00477-018-1605-2.

Knowlton K, Kulkarni SP, Azhar GS, Mavalankar D, Jaiswal A, Connolly M, Nori-Sarma A, Rajiva A, Dutta P, Deol B, Sanchez L, Khosla R, Webster PJ, Toma VE, Sheffield P, Hess JJ, The Ahmedabad Heat and Climate Study Group. Development and implementation of South Asia's first heat-health action plan in Ahmedabad (Gujarat, India). Int J Environ Res Public Health. 2014;11(4):3473–3492. 10. 3390/ijerph110403473.

Kong Q, Huber M. Explicit calculations of Wet-Bulb globe temperature compared with approximations and why it matters for labor productivity. Earth's Future. 2022;10(3):e2021EF002334. https://doi.org/10.1029/2021EF002334.

Kotharkar R, Ghosh A. Review of heat wave studies and related urban policies in South Asia. Urban Clim. 2021;36:100777. https://doi.org/10.1016/j.uclim.2021.100777.

Krishnamurthy M, Ramalingam P, Perumal K, Kamalakannan LP, Chinnadurai J, Shanmugam R, Srinivasan K, Venugopal V. Occupational heat stress impacts on health and productivity in a steel industry in Southern India. Saf Health Work. 2017;8(1):99–104. https://doi.org/10.1016/j.shaw.2016.08.005.

Lu Y-C, Romps DM. Extending the heat index. J Appl Meteorol Climatol. 2022;61(10):1367–83. https://doi.org/10.1175/JAMC-D-22-0021.1.

Lutsko NJ. The relative contributions of temperature and moisture to heat stress changes under warming. J Clim. 2021;34(3):901–17. https://doi.org/10.1175/JCLI-D-20-0262.1.

Mandal R, Joseph S, Sahai AK, Phani R, Dey A, Chattopadhyay R, Pattanaik DR. Real time extended range prediction of heat waves over India. Sci Rep. 2019;9(1):9008. https://doi.org/10.1038/s41598-019-45430-6.

Mazdiyasni O, AghaKouchak A, Davis SJ, Madadgar S, Mehran A, Ragno E, Sadegh M, Sengupta A, Ghosh S, Dhanya CT, Niknejad M. Increasing probability of mortality during Indian heat waves. Sci Adv. 2017;3(6):e1700066. https://doi.org/10.1126/sciadv.1700066.

Monteiro JM, Caballero R. Characterization of extreme Wet-Bulb temperature events in Southern Pakistan. Geophys Res Lett. 2019;46(17–18):10659–68. https://doi.org/10.1029/2019GL084711.

Moore FC, Obradovich N, Lehner F, Baylis P. Rapidly declining remark- ability of temperature anomalies may obscure public perception of climate change. Proc Natl Acad Sci. 2019;116(11):4905–10. https://doi.org/10.1073/pnas.1816541116.

Mukhopadhyay B, Weitz CA. Heat exposure, heat-related symptoms and coping strategies among elderly residents of urban slums and rural villages in West Bengal, India. Int J Environ Res Public Health. 2022;19(19):12446. https://doi.org/10.3390/ijerph191912446.

Murari KK, Ghosh S, Patwardhan A, Daly E, Salvi K. Intensification of future severe heat waves in India and their effect on heat stress and mortality. Reg Environ Chang. 2015;15(4):569–79. https://doi.org/10.1007/s10113-014-0660-6.

Nairn J, Ostendorf B, Bi P. Performance of excess heat factor severity as a global heatwave health impact index. Int J Environ Res Public Health. 2018;15(11) https://doi.org/10.3390/ijerph15112494.

Nori-Sarma A, Anderson GB, Rajiva A, ShahAzhar G, Gupta P, Pednekar MS, Son J-Y, Peng RD, Bell ML. The impact of heat waves on mortality in Northwest India. Environ Res. 2019;176:108546. https://doi.org/10.1016/j.envres.2019.108546.

Normand CWB. The effect of high temperature, humidity, and wind on the human body. Q J R Meteorol Soc. 1920;46(193):1–14. https://doi.org/10.1002/qj.49704619302000004.

Pal JS, Eltahir EAB. Future temperature in southwest Asia projected to exceed a threshold for human adaptability. Nat Clim Chang. 2016;6(2):197–200. https://doi.org/10.1038/nclimate2833.

Perkins-Kirkpatrick SE, Lewis SC. Increasing trends in regional heatwaves. Nat Commun. 2020;11(1):3357. https://doi.org/10.1038/s41467-020-16970-7.

Pillai AV, Dalal T. How is India adapting to heatwaves?: an assessment of heat action plans with insights for transformative climate action (tech. rep.), 2023.

Rao SA, Goswami BN, Sahai AK, Rajagopal EN, Mukhopadhyay P, Rajeevan M, Nayak S, Rathore LS, Shenoi SSC, Ramesh KJ, Nanjundiah RS, Ravichandran M, Mitra AK, Pai DS, Bhowmik SKR, Hazra A, Mahapatra S, Saha SK, Chaudhari HS, et al. Monsoon Mission: a targeted activity to improve monsoon prediction across scales. Bull Am Meteorol Soc. 2019;100(12):2509–32. https://doi.org/10.1175/BAMS-D-17-0330.1.

Raymond C, Matthews T, Horton RM. The emergence of heat and humidity too severe for human tolerance. Sci Adv. 2020;6(19):eaaw1838. https://doi.org/10.1126/sciadv.aaw1838.

Rohini P, Rajeevan M, Srivastava AK. On the variability and increasing trends of heat waves over India. Sci Rep. 2016;6:26153. https://doi.org/10.1038/srep26153.

Rohini P, Rajeevan M, Rao SA, Pillai PA. Assessment of hot weather seasonal temperatures over India using Monsoon Mission Coupled Forecasting System hindcasts. Int J Climatol. 2022;42(16):9823–35. https://doi.org/10.1002/joc.7866.

Romanello M, McGushin A, Napoli CD, Drummond P, Hughes N, Jamart L, Kennard H, Lampard P, Rodriguez BS, Arnell N, Ayeb-Karlsson S, Belesova K, Cai W, Campbell-Lendrum D, Capstick S, Chambers J, Chu L, Ciampi L, Dalin C, et al. The 2021 report of the Lancet Countdown on health and climate change: code red for a healthy future. Lancet. 2021;398(10311):1619–62. https://doi.org/10.1016/S0140-6736(21)01787-6.

Schwalm CR, Glendon S, Duffy PB. RCP8.5 tracks cumulative CO2 emissions. Proc Natl Acad Sci. 2020a;117(33):19656–7. https://doi.org/10.1073/pnas.2007117117.

Schwalm CR, Glendon S, Duffy PB. Reply to Hausfather and Peters: RCP8.5 is neither problematic nor misleading. Proc Natl Acad Sci. 2020b;117(45):27793–4. https://doi.org/10.1073/pnas.2018008117.

Sen B, Dhimal M, Latheef AT, Ghosh U. Climate change: health effects and response in South Asia. BMJ. 2017;j5117 https://doi.org/10.1136/bmj.j5117.

Sherwood SC, Huber M. An adaptability limit to climate change due to heat stress. Proc Natl Acad Sci. 2010;107(21):9552–5. https://doi.org/10.1073/pnas.0913352107.

Simpson CH, Brousse O, Ebi KL, Heaviside C. Commonly used indices disagree about the effect of moisture on heat stress. npj Clim Atmos Sci. 2023;6(1):1–7. https://doi.org/10.1038/s41612-023-00408-0.

Tan CL, Wong NH, Jusuf SK. Effects of vertical greenery on mean radiant temperature in the tropical urban environment. Landsc Urban Plan. 2014;127:52–64. https://doi.org/10.1016/j.landurbplan.2014.04.005.

Taylor KE, Stouffer RJ, Meehl GA. An overview of CMIP5 and the experiment design. Bull Am Meteorol Soc. 2012;93(4):485–98. https://doi.org/10.1175/BAMS-D-11-00094.1.

Vanos JK, Baldwin JW, Jay O, Ebi KL. Simplicity lacks robustness when projecting heat-health outcomes in a changing climate. Nat Commun. 2020;11(1):6079. https://doi.org/10.1038/s41467-020-19994-1.

Venugopal V, Chinnadurai JS, Lucas RAI, Kjellstrom T. Occupational heat stress profiles in selected workplaces in India. Int J Environ Res Public Health. 2016;13(1):89. https://doi.org/10.3390/ijerph13010089.

Venugopal V, Lennqvist R, Latha PK, Suraya A, Jakobsson K, Kjellstrom T. Challenges in conducting epidemiological field studies evaluating associations between heat stress and renal health among workers in informal sectors: experiences from India. Environ Res. 2021a;200:111343. https://doi.org/10.1016/j.envres.2021.111343.

Venugopal V, Shanmugam R, Kamalakkannan LP. Heat-health vulnerabilities in the climate change context—comparing risk profiles between indoor and outdoor workers in developing country settings. Environ Res Lett. 2021b;16(8):085008. https://doi.org/10.1088/1748-9326/ac1469.

Weitz CA, Mukhopadhyay B, Das K. Individually experienced heat stress among elderly residents of an urban slum and rural village in India. Int J Biometeorol. 2022;66(6):1145–62. https://doi.org/10.1007/s00484-022-02264-8.

Wielicki BA, Barkstrom BR, Harrison EF, Lee RB, Louis Smith G, Cooper JE. Clouds and the Earth's Radiant Energy System (CERES): an earth observing system experiment. Bull Am Meteorol Soc. 1996;77(5):853–68. https://doi.org/10.1175/1520-0477(1996)077<0853:CATERE>2.0.CO;2.

Impacts of Climate Change on Public Health in India: An Overview

5

Poornima Prabhakaran

Abstract

Climate change disrupts ecosystems, affecting human health and economies. Human activities exacerbate these impacts by depleting natural resources. The Intergovernmental Panel on Climate Change (IPCC) draws attention to the notable rise in anthropogenic GHG emissions and their role in global warming. Degrading weather patterns and ecosystem disturbances, such as those observed during the COVID-19 pandemic, underscore climate change as a crucial public health challenge. Addressing climate change requires evidence-based global, national, and sub-national policies. Ranked seventh in climate change impact, India faces high social and economic costs from carbon emissions. This chapter explores India's climate change challenges and response initiatives, with particular emphasis on heat, waterborne diseases, nutrition, climate change and conflict, and health systems, including climate-smart healthcare.

Keywords

Climate footprint · Waterborne diseases · Conflict · Climate-smart healthcare · Ecosystems

---

P. Prabhakaran (✉)
Centre for Health Analytics Research and Trends, Trivedi School of BioSciences, Ashoka University, Sonipat, Haryana, India
e-mail: poornima.prabhakaran@ccdcindia.org

H. Brand et al. (eds.), *Climate Change and Public Health*,
https://doi.org/10.1007/978-981-95-0895-2_5

5.1 Introduction

Climate change is disrupting nature's delicate balance across ecosystems and has emerged as a global challenge. The changing weather and environmental factors affect various dimensions including human health and livelihoods, with direct consequences on established and growing economies. Man-made or anthropogenic activities pose an additional hazard through unbridled use of our natural resources like air, water, and land. Recognizing the dire need to protect ecosystems and sustainably manage depleting resources, the UN Decade of Ecosystem Restoration 2021–2030 aims to protect, revive, and restore ecosystems and provide a sustainable future for human beings and nature (UNECE 2023). In its 2007 assessment report, the Intergovernmental Panel on Climate Change (IPCC) draws attention to how the ecosystem is affected by both natural and anthropogenic systems. It also states that 'global GHG[1] emissions due to human activities have grown since pre-industrial times, with an increase of 70 percent between 1970 and 2004', and that 'most of the observed increase in global average temperatures since the mid-20th century is very likely due to the observed increase in anthropogenic GHG concentrations' (Dickson et al. 2021).

Changing weather patterns, particularly in recent years, have led to the degradation of ecosystems. The consequent blurring of boundaries between human, animal, and plant systems has caused changes in natural resources and even built environments leading to both the depletion and pollution of those limited natural resources. The recent global COVID-19 pandemic exemplified the disastrous consequences of ecosystem disturbances. The insidious onset of the other global pandemic—climate change—is increasingly recognized as the biggest challenge to public health. The growing global attention to climate change presents the biggest opportunity to mount a suitable response to protect population health. Global, national, and sub-national programs and policies to address climate change must be informed by contextualized research and evidence on climate change impacts on public health (WHO 2021).

Various initiatives have been undertaken by global actors to curb climate change (Climate champions 2022; Maizland 2022). Other chapters in this book delve deeper into these actions at the global level. At a country level, because of its diverse climate zones, topography, and ecosystems, India is especially vulnerable to large-scale climate change-related risks. According to the Global Climate Risk Index 2021, India is ranked as the seventh most climate change-affected country in the world (Eckstein et al. 2021; Sharma 2023). A World Bank report on South Asia's hotspots reported India amongst the countries who most experience high social and economic costs from climate change: each additional ton of carbon dioxide emission costs India USD 86, followed by the USA at USD 48, and Saudi Arabia at USD 47.5 (Mani et al. 2018). The multifaceted impacts of climate change on India require a concerted response from all sectors and driven by local evidence. This chapter looks at the intersection of climate change and public

[1] Greenhouse gas

health and examines the efforts and initiatives undertaken by India to mount a suitable response.

5.2 Emerging Patterns Due to Climate Change

The Lancet Report (2020) revealed that most industrialized countries were responsible for 90 percent of excess global CO_2 emissions in contrast to developing countries. The USA was responsible for 40 percent of emissions as of 2015 followed by the EU at 29 percent (Hickel 2020). National fair shares of a safe global carbon budget consistent with the planetary boundary of 350 ppm were derived. These fair shares were then subtracted from countries' actual historical emissions (territorial emissions from 1850 to 1969, and consumption-based emissions from 1970 to 2015). Its findings drew attention to the climate change damages suffered by low-income countries despite not being major contributors to the emissions (Hickel 2020). However, given the deleterious impacts, particularly on health, all nations need to take action to curb emissions and strengthen their health system response.

Climate change-related risks have multidimensional consequences. Changing conditions in temperature, rainfall, and sea levels that are responsible for acute climate events such as heat waves, floods, storms, droughts, and deteriorating air quality impact public health (Horton 2009; Levy and Patz 2015).

Climate change exacerbates social inequalities, disproportionately affecting marginalized groups and low-income nations. It leads to severe environmental and health crises, such as extreme weather events (droughts, increased sea level, increased temperature, etc.), reduced agricultural yields, displacement (environmental refugees), and the spread of diseases including vector-borne and water diseases, and mental health illnesses. These challenges threaten basic human rights, including access to food, water, healthcare, and security. Vulnerable groups include individuals especially in low-income regions and disadvantaged groups who bear the brunt of these effects, although they contribute the least to greenhouse gas emissions (GHGs) (Horton 2009).

5.2.1 Extreme Heat

Between 1901 and 2007, there was a 0.5-degree increase in mean temperature in India, with regional variations and a potential for a further 2.2–5.5-degree rise in northern, central, and western parts of the country (Prabhakaran and Reddy 2018). Such a steady increase in temperature can cause direct and indirect consequences on population health.

Dense urban and commercial pockets of India also experience urban heat island (UHI) intensities of up to 8–10 °C (Mohan et al. 2022). India is now the most populous country in the world with increasing urbanization and associated concretization. The concrete infrastructure absorbs and re-emits the sun's heat as compared to natural elements such as forests and compounds the effects of rising temperatures. According to a World Bank Report (2018), it 'could cost India 2.8 percent of GDP and depress the living standards of nearly half the country's population by 2050' (Mani et al. 2018; The World Bank 2018). A recent study by the Lancet stated that extreme heat led to a 55 percent increase in heat-related deaths in India from 2000 to 2004 and 2017 to 2021. It also resulted in income losses that were roughly 5.4 percent of the nation's GDP (BBC 2022; Golechha and Panigrahy 2020; WMO 2022). In March–May 2022, India and Pakistan recorded the highest land temperatures that caused power failures, crop damage, wildfires, and heat-related deaths (The Lancet Planetary Health 2022).

Extreme heat can cause a range of health impacts from simple heat exhaustion, muscle cramps, and syncope to dehydration, kidney failure, and exacerbation of cardiovascular and respiratory illnesses. Infants, children, pregnant women, and elderly individuals are particularly vulnerable besides occupational groups with long outdoor working hours like farmers, traffic police, and daily-wage labourers. Health professionals must be equipped to both recognize and respond to heat-related illnesses encountered in clinical practice through preventive, promotive, and curative strategies (Sorensen and Hess 2022).

Through the National Disaster Management Authority, the government of India has established a national framework for heat action plans to manage rising temperatures. The National Action Plan for Climate Change and Human Health commissioned by the Ministry of Health and Family Welfare in 2018 also developed a Heat Adaptation Plan for state and district-wide adaptation response. Ahmedabad in Gujarat, India, experienced a devastating heat wave in 2010 with temperatures touching 47–48 degrees Celsius and 800 excess deaths occurring within a single week from 20 to 27 May (Fig. 5.1) (Azhar et al. 2014). The city officials with public health specialists developed and implemented a city-wide heat health adaptation plan in the following years (WMO 2022) convening cross-sectoral actors including public health practitioners, exposure scientists, the state meteorological department, and city-level actors including urban local body representatives, municipal bodies, traffic police, and school authorities. This inter-sectoral coordinated response to address a heat wave by preventive actions has been successfully shown to reduce mortality in the city year after year. The Ahmedabad Heat Action Plan (AHAP) has provided a successful template for other cities across India. Odisha, another heat wave-prone state with spikes in mortality, also has its heat action plan. With the number, frequency, and duration of heat waves steadily increasing across other parts of India, the Ministry of Health in India ramped up efforts through the National Program for Climate Change and Human Health to stimulate countrywide actions. The heat action plans are being scaled up across the board to include 23 heat wave-prone states and address more than 130 cities and districts (WMO 2022).

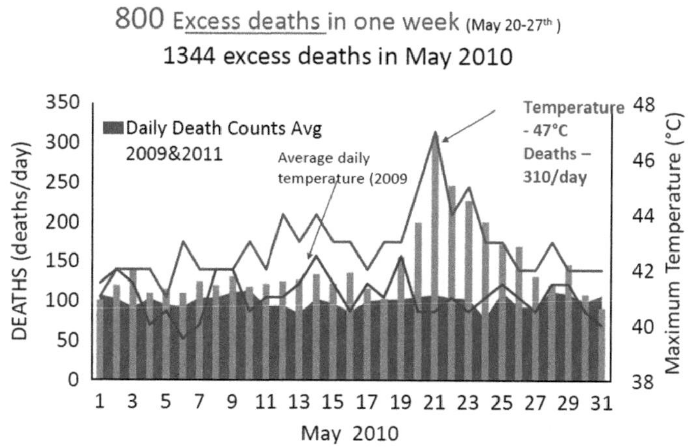

Fig. 5.1 Data on the number of deaths in the Ahmedabad heat wave (2010). (Adapted from Azhar et al. 2014)

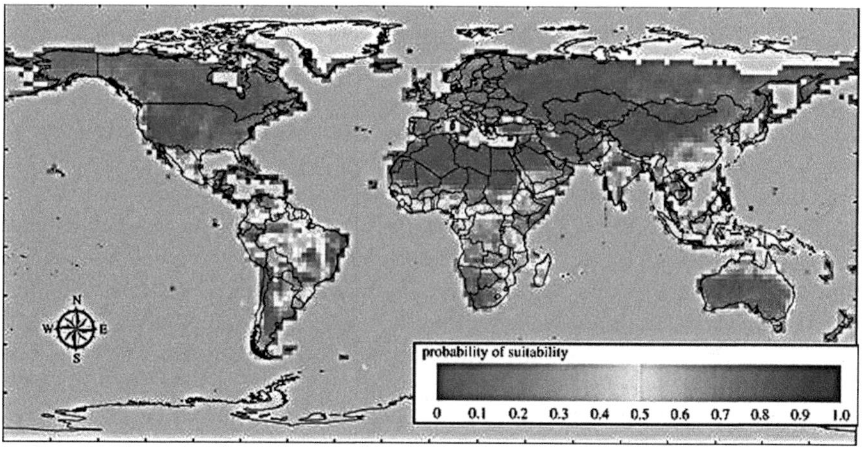

Fig. 5.2 Global change in VBD patterns. (By Rogers 2015)

5.2.2 Change in Vector-Borne Disease Patterns

Changes in climatic conditions have resulted in creating favourable conditions for the spread of vector-borne diseases (VBD). Globally, the hotspots for VBDs are increasing, particularly for mosquito-borne diseases (Fig. 5.2) (Rogers 2015). In the case of dengue, the hot spots are mostly in Southeast Asia as well as in Latin America (CDC 2021), with India harbouring a huge burden of this disease. Within the country, the VBD patterns are steadily changing with diseases spreading to the northeastern states of India, beyond traditional hotspots. Favourable climatic conditions spur both the spread of the vector and decrease in the extrinsic incubation period of the mosquito resulting in the widespread dissemination of dengue even in hilly areas (Fig. 5.3). Likewise, malaria is now spreading in India to all the states in the north and northeast, including Arunachal Pradesh, Nagaland, Manipur, and Mizoram, where the disease was not widely seen. There is a longer transmission window, and the transmission period is not restricted to the traditional monsoon seasons known to favour vector-borne mosquito diseases. It has widened between 10 and 12 months in cases of malaria. Chikungunya is another mosquito-borne disease that has shown changing patterns because of changing climatic conditions. States again in the Northeast not previously exposed are beginning to see more cases of this debilitating disease (Dhiman et al. 2010; Singh et al. 2011).

5.2.3 Waterborne Diseases

Changing precipitation patterns in recent years have led to unseasonal rainfall in many Indian states. Rising temperatures and sea water levels have led to several cyclonic events and floods in coastal areas like Kerala, Tamil Nadu, Orissa, and Gujarat with states like Assam, Mumbai, and now even Delhi and the National Capital Region facing urban floods with alarming frequency.

The post-flood impacts on health and health systems have been particularly devastating over the last 5–6 years. Storm water overflow and the resulting waterborne diseases is a major public health concern in a country like India where the burden of communicable diseases has continued unabated even as both rural and urban areas see a surge in non-communicable diseases. The disruption of healthcare in post-flood situations further worsens the coping capacity to deal with the disease burden thereby increasing the propensity for illness and death, particularly in vulnerable groups. Additionally, India is among the water-stressed countries of the world with less than 1700 m^3/person/year of water at its disposal. In the contrasting situation of droughts following rising temperatures and heat waves, the limited sources of water, often polluted, further fuel the onset of disease. More than 600 million people in India are estimated to be facing acute water shortages (Chattopadhyay 2022). There is, therefore, an urgent need to optimize water management systems at both health facility and community levels, establish rainwater harvesting systems where surplus rainfall can be leveraged for lean periods of water scarcity and availability, and strengthen the health system response to water-related illnesses.

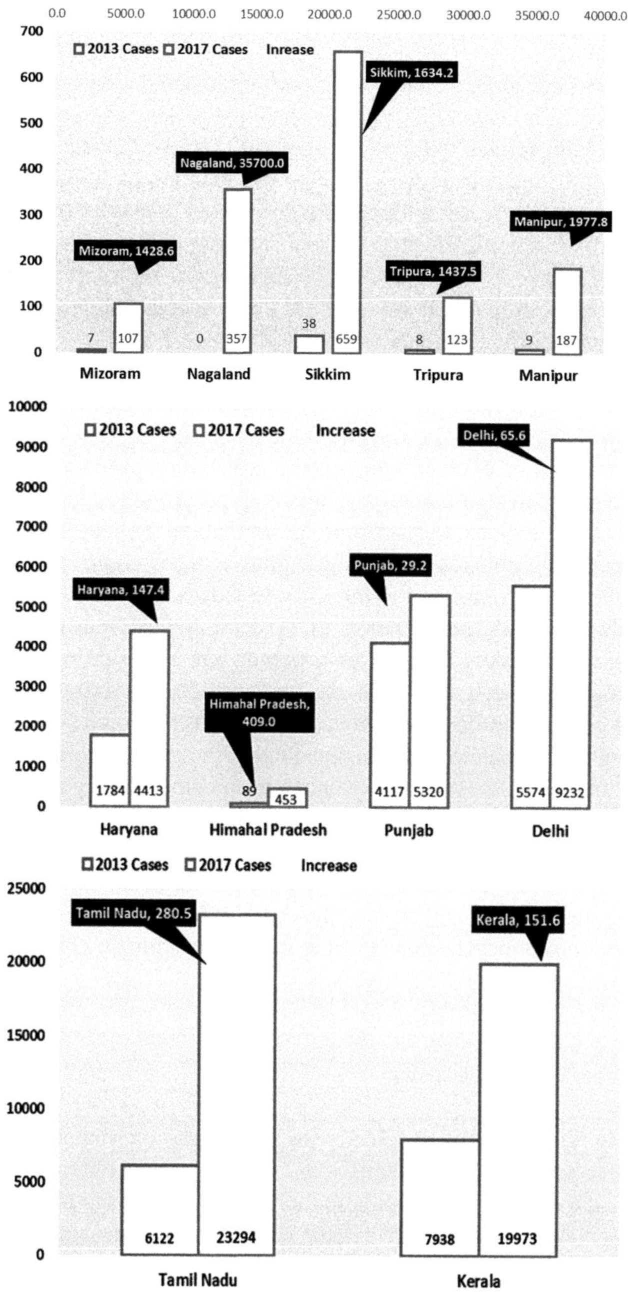

Fig. 5.3 VBD (Dengue) patterns across India. (By Mutheneni et al. 2017)

5.3 Through the Nutrition Lens

India is an agrarian economy, and a large part of the population is dependent on the agriculture sector for livelihood. With the changing climate and global warming, 40 percent of the land surface will be experiencing altered climates (Suri and Sharma 2022). The direct impact in India will be on the quantum of crops produced with a two percent annual decrease in the annual production of staple crops—rice, wheat, maize, sorghum, and bajra. This has consequences for food security even as increasing temperature impacts the rates of crop maturation. With rapid maturation, the nutrient content of crops is at sub-optimal levels thereby adding a propensity to worsen nutrient security (Bandara and Cai 2014).

Using a global dynamic computable general equilibrium model, how climate change-induced shifts in crop productivity influence food prices and food security in South Asia was analysed, focusing on Bangladesh, India, Nepal, Pakistan, and Sri Lanka. The findings reveal that reductions in agricultural productivity declines due to climate change will significantly negatively impact food production and increase prices across these nations. Furthermore, given that nearly half of the world's poor population resides in this region and agriculture is a key contributor to GDP, food security challenges are expected to intensify (Bandara and Cai 2014).

The country is still in the phase of epidemiological transition with under-nutrition-related disorders as well as a population transition towards processed foods fuelling the other extreme of the malnutrition spectrum—overweight and obesity. The growing burden from three simultaneous pandemics—under-nutrition, overnutrition, and climate change—is labelled as the global 'syndemic' (Swinburn et al. 2019), including obesity, under-nutrition, and other dietary risks, which is the leading cause of poor health globally. In the near future, the health effects of climate change will considerably compound these health challenges. Climate change can be considered a pandemic because of its sweeping effects on the health of humans and the natural systems we depend on (i.e. planetary health necessitating a multipronged health system and policy response to the impacts of climate change on food and food systems).

5.4 Livestock Breeding

The global livestock sector specifically the ruminants contributes to 32 percent of the total anthropogenic methane emissions. These are projected to increase up to 70 percent by 2050 due to rising population, urban migration, and economic development (Johnson and Johnson 1995; UNEP 2021). In India, the agricultural sector is responsible for total emissions of methane (CH4) and nitrous oxide (N2O) of 10214.8 Gg and 0.07 Gg, respectively. Around 63 percent of all emissions from India's agriculture are attributable to livestock (Mosier et al. 1998; Sharma 2020). Comparing other agricultural subsectors, such as rice cultivation and open burning of crop residue, Indian livestock contributes the most methane emissions (Sharma 2020). While this warrants attention, the policy response must also address the issue from a social lens given the livelihoods issue for a large sector of the small- and medium-scale farming households hosting livestock.

5.5 Climate Change and Conflict

Acute climate-related events often result in the movement of populations from their original homes and habitats. India has seen several waves of migration in recent years due to acute climate events, particularly from flood threats in coastal areas and erratic rainfall patterns resulting in urban flooding and population movement in several regions (Chaudhury and Sen 2022).

While temporary climate-related migration can lead to diminished resources for necessities like food, water, housing, and even healthcare, long-term stress situations can also precipitate other issues. For example, lower availability and limited resources can lead to conflict situations in terms of access to food, access to homes, and livelihoods. Conflict risk could increase because of water insecurity. Drought-related increases in food prices can exacerbate disputes and encourage both conflict and migration. Periods of drought and flooding have led to waves of migration and an increase in violence within countries thereby affecting economic growth (The World Bank 2023). Sustained climate risk can impact opportunities for access to education and healthcare thereby leading to long-term impacts for current and future generations. Climate-related migration and climate refugees therefore need focussed efforts for acute care during crises and long-term rehabilitation. In the current times, every one-degree increase in temperature can increase the frequency of interpersonal conflict by about 14 percent (Hsiang et al. 2013; IPCC 2018).

5.6 Climate Change and Its Impacts on Mental Health

The WHO policy brief (2022) launched at the Stockholm+50 conference draws attention to the potential serious risks climate change poses to mental health and well-being. Thus, the WHO is urging nations to incorporate mental health assistance into their response to the climate catastrophe (WHO 2022). Rising temperatures are anticipated to increase violent suicides and rates of violence, while extended droughts brought on by climate change may increase the number of farmer suicides. Droughts can result in stress and poor mental health. Depression, adjustment disorders, and post-traumatic stress disorder can develop as a result of the increased frequency of disasters brought on by climate change (Padhy et al. 2015; Palinkas and Wong 2020). The range of impacts has not been well understood. This review examines mental health impacts of three types of climate-related events: 1. Youth are extremely sensitive to the impacts of climate change on their future, and 'solastalgia' or perceived distress due to environmental risks is particularly evident amongst the younger age groups. Eco-anxiety or a chronic fear of impending environmental doom is increasingly an area of concern amongst youth. Focussed efforts need to be directed to address climate impacts on the mental health of vulnerable groups like children, women, and the elderly besides poorer socio-economic groups. However, India's mental health infrastructure remains severely lacking with very few qualified mental health practitioners, variable quality of care, low levels of investment, and societal stigma associated with mental health. The compounding mental health effects of climate change pose an additional challenge that India is ill-equipped for.

5.7 Climate Change and the Health System

5.7.1 Sectoral Emissions

The health sector, under its activities, also contributes to GHG emissions. If all the health facilities and health systems around the world were put together and ranked with countries, the health sector would be the fifth largest emitter on the planet. Among the top five emitters are the United States, China, the European Union, Japan, and Russia, with India, Brazil, South Korea, Canada, and Australia contributing to two percent of those emissions. All other countries of the world are contributing to 25 percent of the emissions from the health sector.

5.8 India's Healthcare Climate Footprint

According to a European Parliament report, India accounted for seven percent of global GHG emissions in 2019 (Jensen and Delivorias 2022). India accounted for 7% of global GHG emissions; its per capita emissions level of 2.48 tonnes CO2 equivalent (tCO2e). As per a report from the international organization Healthcare Without Harm 2019, the Indian health sector contributes about 39 metric tons of carbon dioxide, equivalent to about ten coal-fired power plant emissions in one year (Karliner et al. 2019). While healthcare service delivery and hospital operations like the use of building infrastructure, energy, water, and waste generation and disposal contribute some proportion of the hospital climate footprint, a larger proportion arises from the supply chain. Therefore, there is immense scope to address the adoption of sustainable practices in the production, supply use, and disposal of goods and services used in the health sector like food, drugs, pharmaceuticals, and medical equipment. Health professionals must be sensitized and equipped with the knowledge and resources to handle the anticipated surge in disease burden. While health systems hold primary responsibility to deliver uninterrupted health services even in crises, strengthening our health infrastructure and workforce must go in tandem with building environmentally friendly health facilities. The dual role of building an adaptation response and structural and functional resilience with emissions reduction will help propel the establishment of climate–smart health systems globally and in India. This issue is dealt with greater detail in a later chapter. Suffice to say here that ongoing efforts to achieve universal health coverage must simultaneously strike a balance between facilitating equitable and affordable access as well as environment-friendly and sustainable health facilities.

The 2015 Paris Agreement, which called for aligning healthcare action and mobilizing the health sector as a leader in responding to the climate crisis, is a clarion call to health actors to address climate change as frontline and 'first responders'.

5.9 Climate Smart Healthcare

Healthcare Without Harm and the World Bank in 2017 guided seven different dimensions of climate-smart healthcare (World Bank 2017). This provided insights into the overall system design of health systems, with careful attention to building design and construction, investment in renewable energy, and energy efficiency in health systems. It also emphasized optimal waste management, sustainable transport, and water consumption policies in healthcare facilities, along with sustainable procurement of pharmaceuticals, medical devices, food, and other products, as well as overall resilience strategies to address the increasing disease burden. In its guidance document in 2020 October, WHO guides the development of climate-resilient and environmentally sustainable healthcare facilities. This guidance focusses on four areas for interventions—health workforce capacity building, water, sanitation, hygiene and healthcare waste management, and sustainable energy services—but also focuses on infrastructure, technologies, and products that are sustainable, ensuring that the health systems are both structurally resilient and functionally capacitated to deal with the impacts of climate change (WHO 2020).

The Ministry of Health and Family Welfare in India's Health Adaptation framework developed in collaboration with its Centre of Excellence at the Public Health Foundation of India provides implementation guidelines to both the public and private health facilities in India to transition to climate-resilient and environmentally friendly health facilities. The framework advocates an incremental approach based on available resources to adopt the domain-specific interventions for optimal energy and water management besides addressing other pillars of action like sustainable building infrastructure, waste management, procurement, transport, and linkages to climate information services for early warning systems. The baseline conducts of climate vulnerabilities at regional and health system levels are essential first steps that will help health facilities develop their time-sensitive and customized climate action plans. The commitment of health sector leadership and climate finances is an important prerequisite to ensure the implementation of the national adaptation plan in India to address climate change and its impacts on human health.

With several countries committing to decarbonize their health systems at the Glasgow COP 26 conference and a follow-up effort by the WHO through its Alliance for Transformative Action on Climate and Health (ATACH), the needle will move at the global level (Karliner et al. 2023). At the G-20 health sector working group meeting hosted by India in April 2023, a new set of climate resilience principles was released for consideration and action by G-20 countries (Asian Development Bank, G20 Secretariat 2023). A policy brief released after this outlines a roadmap for global healthcare decarbonization with a strategy for India to adopt a late peak strategy conforming to the country's needs to still focus on universal health coverage even as it mounts a suitable adaptive and mitigation response for its diverse health system (Prabhakaran et al. 2023). At the core of this effort, however, is the need for convergence of all important stakeholders and climate-centric programs at national and sub-national levels so that inter-sectoral coordination can stimulate decentralized simultaneous and coherent responses to mitigate the climate crisis.

References

Asian Development Bank, G20 Secretariat, G. of I. M. of H. and F. W. Briefing document: addressing the challenges of climate change and health, one earth, one family, one future. Asian Development Bank; 2023.

Azhar GS, Mavalankar D, Nori-Sarma A, Rajiva A, Dutta P, Jaiswal A, Sheffield P, Knowlton K, Hess JJ. Heat-related mortality in India: excess all-cause mortality associated with the 2010 Ahmedabad heat wave. PLoS One. 2014;9(3) https://doi.org/10.1371/JOURNAL.PONE.0091831.

Bandara JS, Cai Y. The impact of climate change on food crop productivity, food prices and food security in South Asia. Econom Anal Policy. 2014;44(4):451–65. https://doi.org/10.1016/j.eap.2014.09.005.

BBC. India heatwave: High temperatures killing more Indians now, Lancet study finds - BBC News, 2022, October 26. https://www.bbc.com/news/world-asia-india-63384167.

CDC. Areas with Risk of Dengue | Dengue | CDC, 2021, November 5. https://www.cdc.gov/dengue/areaswithrisk/index.html.

Chattopadhyay S. India's growing water crisis, the seen and the unseen - The Hindu. The Hindu, 2022, September 15. https://www.thehindu.com/opinion/op-ed/indias-growing-water-crisis-the-seen-and-the-unseen/article65891233.ece.

Chaudhury ABR, Sen P. Climate change fueling climate migration | ORF. Observer Research Foundation, 2022, December 19. https://www.orfonline.org/expert-speak/climate-change-fueling-climate-migration/.

Climatechampions. UN climate change high-level champions: contribution of the 'All of Society' Global Climate Action Agenda at COP27 - Climate Champions, 2022, November 17. https://climatechampions.unfccc.int/contribution-of-the-all-of-society-global-climate-action-agenda-at-cop27/

Dhiman RC, Pahwa S, Dhillon GPS, Dash AP. Climate change and threat of vector-borne diseases in India: are we prepared? Parasitol Res. 2010;106(4):763–73. https://doi.org/10.1007/s00436-010-1767-4.

Dickson B, Miles L, Thornton H, O'Connell E. Ecosystem restoration for people, nature and climate becoming #GenerationRestoration. United Nations Environment Programme, 2021. http://www.un.org/Depts/Cartographic/.

Eckstein D, Vera K, Schäfer L. Global climate risk index 2021 | Germanwatch e.V. Germanwatch, 2021, January. https://www.germanwatch.org/en/19777.

Golechha M, Panigrahy RK. COVID-19 and heatwaves: a double whammy for Indian cities. Lancet Planetary Health. 2020;4(8):e315–6. https://doi.org/10.1016/S2542-5196(20)30170-4.

Hickel J. Quantifying national responsibility for climate breakdown: an equality-based attribution approach for carbon dioxide emissions in excess of the planetary boundary. Lancet Planetary Health. 2020;4(9):e399–404. https://doi.org/10.1016/S2542-5196(20)30196-0.

Horton R. The climate dividend. Lancet. 2009;374(9705):1869–70. https://doi.org/10.1016/S0140-6736(09)61994-2.

Hsiang SM, Burke M, Miguel E. Quantifying the influence of climate on human conflict. Science. 2013;341(6151) https://doi.org/10.1126/SCIENCE.1235367/SUPPL_FILE/HSIANG.SM.PDF.

IPCC. Impacts of 1.5°Cof global warming on natural and human systems. In: Masson-Delmotte V, Zhai P, Pörtner H-O, Roberts D, Skea J, Shukla PR, Pirani A, Moufouma-Okia W, Péan C, Pidcock R, Connors S, Matthews JBR, Chen Y, Zhou X, Gomis MI, Lonnoy E, Maycock T, Tignor M, Waterfield T, editors. Global warming of 1.5°C. An IPCC special report. Cambridge University Press; 2018. p. 175–312. https://doi.org/10.1017/9781009157940.005.

Jensen L, Delivorias A. International progress on climate action India's climate change policies State of play ahead of COP27. European Parliamentary Research Service; 2022.

Johnson KA, Johnson DE. Methane emissions from cattle. J Anim Sci. 1995;73(8):2483–92. https://doi.org/10.2527/1995.7382483x.

Karliner J, Slotterback S, Boyd R, Ashby B, Steele K. Health care's climate footprint how the health sector contributes to the global climate crisis and opportunities for action. Health Care Without Harm. 2019:1–48.

Karliner J, Osewe P, Neira M, Arora D, Galvao L, Reddy KS. Momentum builds for health-care climate action. Lancet. 2023;0(0) https://doi.org/10.1016/s0140-6736(23)01079-6.

Levy BS, Patz JA. Climate change, human rights, and social justice. Ann Glob Health. 2015;81(3):310–22. https://doi.org/10.1016/j.aogh.2015.08.008.

Maizland L. Global climate agreements: successes and failures | Council on Foreign Relations, 2022, November 4. https://www.cfr.org/backgrounder/paris-global-climate-change-agreements.

Mani M, Bandyopadhyay S, Chonabayashi S, Markandya A, Mosier T. South Asia' s hotspots the impact of temperature and precipitation changes on living standards. World Bank; 2018. p. 13–125. https://doi.org/10.1596/978-1-4648-1155-5.

Mohan M, Bhati S, Sati AP. Urban heat Island effect in India: assessment, impacts, and mitigation. Global Urban Heat Island Mitig. 2022:199–250. https://doi.org/10.1016/B978-0-323-85539-6.00007-X.

Mosier AR, Duxbury JM, Freney JR, Heinemeyer O, Minami K. Assessing and mitigating N2O emissions from agricultural soils. Clim Chang. 1998;40(1):7–38. https://doi.org/10.1023/A:1005386614431.

Mutheneni SR, Morse AP, Caminade C, Upadhyayula SM. Dengue burden in India: recent trends and importance of climatic parameters. Emerg Microb Infect. 2017;6(8):1–10. https://doi.org/10.1038/EMI.2017.57.

Padhy SK, Sarkar S, Panigrahi M, Paul S. Mental health effects of climate change. Indian J Occup Environ Med. 2015;19(1):3. https://doi.org/10.4103/0019-5278.156997.

Palinkas LA, Wong M. Global climate change and mental health. Curr Opin Psychol. 2020;32:12–6. https://doi.org/10.1016/j.copsyc.2019.06.023.

Prabhakaran P, Reddy S. Briefing for Indian policymakers. Lancet. 2018:1–20.

Prabhakaran P, Armstrong F, Karliner J. Decarbonising the healthcare sector: a roadmap for G20 countries. Observer Research Foundation, 2023, pp. 3–22. https://www.orfonline.org/wp-content/uploads/2023/06/T20_PolicyBrief_TF3_DecarbonisingHealthcare.pdf.

Rogers DJ. Dengue: recent past and future threats. Philos Trans R Soc B Biol Sci. 2015;370(1665):1–18. https://doi.org/10.1098/RSTB.2013.0562.

Sharma UC. Methane and nitrous oxide emissions from livestock in India: impact of land use change. J Agric Aquacult. 2020, March 19;2(1) https://escientificpublishers.com/methane-and-nitrous-oxide-emissions-from-livestock-in-india-impact-of-land-use-change-JAA-02-0014

Sharma S. India's climate change policy: challenges and recommendations - ISPP. Policy Review, 2023, January 16. https://www.ispp.org.in/policyreview_blog/indias-climate-change-policy-challenges-and-recommendations/.

Singh BB, Sharma R, Gill JPS, Aulakh RS, Banga HS. Climate change, zoonoses and India. Revue Scientifique et Technique (International Office of Epizootics). 2011;30(3):779–88. https://doi.org/10.20506/rst.30.3.2073.

Sorensen C, Hess J. Treatment and prevention of heat-related illness. N Engl J Med. 2022;387(15):1404–13. https://doi.org/10.1056/NEJMCP2210623.

Suri S, Sharma N. Climate change poses the biggest risk to food security in India | ORF. Observer Research Foundation, 2022, August 29. https://www.orfonline.org/expert-speak/climate-change-poses-the-biggest-risk-to-food-security-in-india/

Swinburn BA, Kraak VI, Allender S, Atkins VJ, Baker PI, Bogard JR, Brinsden H, Calvillo A, De Schutter O, Devarajan R, Ezzati M, Friel S, Goenka S, Hammond RA, Hastings G, Hawkes C, Herrero M, Hovmand PS, Howden M, et al. The global syndemic of obesity, undernutrition, and climate change: the Lancet Commission report. Lancet. 2019;393(10173):791–846. https://doi.org/10.1016/S0140-6736(18)32822-8/ATTACHMENT/4213811F-27DE-4ECF-9474-CE263B58DA4F/MMC1.PDF.

The Lancet Planetary Health. Be prepared for more heat. Lancet Planetary Health. 2022;6(9):e706. https://doi.org/10.1016/S2542-5196(22)00201-7.

The World Bank. Climate change could depress living standards in India, says New World Bank Report, 2018, June 28. https://www.worldbank.org/en/news/press-release/2018/06/28/climate-change-depress-living-standards-india-says-new-world-bank-report.

The World Bank. High and dry: climate change, water, and the economy. World Bank, 2023. https://www.worldbank.org/en/topic/water/publication/high-and-dry-climate-change-water-and-the-economy.

UNECE. UN decade on ecosystem restoration | UNECE. United Nations Economic Commission for Europe, 2023. https://unece.org/un-decade-ecosystem-restoration.

UNEP. Methane emissions are driving climate change. Here's how to reduce them. UN Environment Programme, 2021, August 20. https://www.unep.org/news-and-stories/story/methane-emissions-are-driving-climate-change-heres-how-reduce-them.

WHO. WHO guidance for climate resilient and environmentally sustainable health care facilities. World Health Organization, 2020, October 12. https://www.who.int/publications/i/item/climate-resilient-and-environmentally-sustainable-health-care-facilities.

WHO. Climate change and health research: current trends, gaps and perspectives for the future. World Health Organization; 2021. p. 1–48. https://cdn.who.int/media/docs/default-source/climate-change/cop26/final_cop26-research-web.pdf?sfvrsn=e8fb5dcb_5&download=true

WHO. Why mental health is a priority for action on climate change. World Health Organization, 2022, June 3. https://www.who.int/news/item/03-06-2022-why-mental-health-is-a-priority-for-action-on-climate-change.

WMO. Climate change made heatwaves in India and Pakistan "30 times more likely" | World Meteorological Organization, 2022. https://public.wmo.int/en/media/news/climate-change-made-heatwaves-india-and-pakistan-30-times-more-likely.

World Bank. Low-carbon and resilience strategies for the health sector. World Bank, 2017. www.worldbank.org.

Impact of Climate on Health and Governance in India

6

Atul Kotwal and Taruna Juneja Gandhi

Abstract

Climate change is a global crisis with multi-scalar characteristics affecting populations. Despite global efforts, the goals remain unmet, with developed countries failing to meet their commitments. India has prioritized climate change on a global scale, aiming for net zero by 2070. Major health impacts include emerging infections, mental distress, and non-communicable diseases. This chapter discusses the socio-economic and environmental determinants of health affected by climate change, including biodiversity, deforestation, and natural disasters. It discusses India's national-level policies on climate change and health, including the Government of India's initiatives towards environmentally sustainable healthcare and climate-smart health systems. It also highlights state action plans on climate change and the inclusion of health as a component. It emphasizes the need for a comprehensive plan to combat climate change's effects on health, focusing on mitigation, adaptation, and the resilience of vulnerable populations.

Keywords

Climate policy · Health policy · India · Climate footprint · Health systems resilience

A. Kotwal (✉) · T. J. Gandhi
National Health Systems Resource Centre, Ministry of Health, Government of India, New Delhi, India
e-mail: atul.kotwal@nhsrcindia.org

H. Brand et al. (eds.), *Climate Change and Public Health*,
https://doi.org/10.1007/978-981-95-0895-2_6

6.1 Introduction

Climate change is one of the most serious global environmental crises that is unique in its multi-scalar characteristics from global to local. It is becoming an existential problem and will eventually affect the entire population across the globe. Despite numerous efforts globally in the form of consultations, conferences, and discussions, the results are yet unimpressive. The developed world has failed to keep its promises towards climate change in comparison to its wealth and capacities. Despite the unrealistic expectations from developing countries, India has prioritized the climate change agenda on a global scale by addressing development paradigms while ensuring growth and development of the economy along low carbon pathways towards net zero by 2070 (Ministry of Environment, Forest and Climate Change, Government of India 2022). Considering the effects of climate change being felt and noticeable globally, developing countries like India have stressed the need to address the impact on global health through addressing paradigms of development focussing on human well-being; human resources; and social, inclusive, and sustainable development. As this is an issue affecting the entire world, India in the true spirit of '*Vasudhaive Kutumbkam*', has been and will put in concerted efforts to limit climate change and its effects. However, the targets which have been given need further discussion in a larger forum. There is a felt need for an overarching plan and framework to combat the effects of climate change on the health of the population at large. Starting with the factors responsible, highlighting vulnerable populations, development of an effective mitigation plan, and its effective implementation—which is agreed by countries across the globe.

Climate change affects the socio-economic and environmental determinants of health. It has various adverse effects on biodiversity; deforestation; habitat; and ocean and marine health. Other challenges include disasters—floods, droughts, heatwaves, storms; shifting seasons; and extreme temperatures. There are significant harmful effects of climate change on the health of the population, which include but are not limited to, melting permafrost exposing frozen viruses, emerging and re-emerging infections, AMR, mental distress, increase in certain infectious diseases like diarrhoea, and increase in NCDs like cancers and kidney diseases (CDC 2022; Patz et al. 2005). The short- to medium-term impact of climate change is determined by the magnitude and rate of change, vulnerability of the populations, their resilience to the rate of climate change, pace of adaptation, etc. In the long term, the health impact of climate change depends on the actions taken to counter its impact; measures taken to reduce the emissions; to avoid reaching dangerous temperature thresholds; and to avoid potential irreversible tipping points.

This chapter will outline the interlinkages and connection between climate change and health systems, followed by the Government of India's (GoI) plans and programs towards climate change and environmentally sustainable healthcare facilities as well as the resources allocated for the same. This chapter will highlight what climate-smart healthcare and health systems are and describe India's various initiatives to reduce the impact of climate change on health and how they are linked (NHSRC 2023c).

6.2 How Are Climate Change and Health Systems Connected?

Climate change and health systems are connected in a variety of ways. One, there is a growing health threat from climate change as seen through a rising number of weather events, accidents, injuries, VBD, zoonotic diseases, and emerging and re-emerging infections. Secondly, health systems are also a part of the problem because 4.4 percent of the global GHG emissions are by the health sector (Karliner et al. 2020). Therefore, the health sector is both the contributor and the sufferer because of climate change. Importantly, the current health systems are inadequately equipped to manage these changing health risks. There are many developed countries without much incidence/prevalence of VBD or communicable diseases currently; however, they will face challenges in the management of these diseases or will require extra resources for the management of these diseases.

Factors that increase the vulnerability towards climate change can be demographic, geographic, biological, health status, socio-political, and socio-economic conditions. Exposure pathways include extreme weather, heat stress, air quality, water quality and quantity, food security and safety, vector distribution, and ecology. Linking these with health system outcomes and developing 'resilience' is of utmost importance now, which can only be addressed by strengthening governance and leadership.

India's healthcare climate footprint falls into three major blocks. The first is healthcare facilities themselves, including vehicles used in the healthcare industry. Immediate actions can be taken to reduce healthcare facility emissions. Second, are the energy sources such as electricity, wherein there is a need to invest and educate for decarbonization of local and national energy systems and implementation of clean renewable energies. The third one is the supply chain, which consists of the production and procurement of equipment, along with the disposal of medicines, drugs, consumables, and other materials. For this, we need to set up and implement criteria for low-carbon or zero-emissions procurement to decarbonize the supply chain. Figure 6.1 depicts India's healthcare climate footprint—Scope 1 is eight percent by the healthcare facilities themselves and the vehicles being used; followed by Scope 2 which is 11 percent consisting of energy sources; and Scope 3 forms the major portion with 81 percent by the supply chain (Karliner et al. 2020).

From the healthcare system's point of view, there is a need to invest in climate-resilient and environmentally sustainable healthcare facilities. Quoting an example from India: in 2018, Kerala witnessed floods that tremendously impacted its public health system (World Health Organization 2020). The hospitals faced power outages for 3–9 days causing the shutdown of cold storage systems. There was damage to the entire stock of vaccines and other refrigeration-dependent medical supplies, and IT equipment was impacted. The inability to cater to patients, especially during emergencies, was another important loss that was much worse than the financial loss.

The 2012 Hurricane Sandy in the USA forced the evacuation of more than 6400 patients from six hospitals, and New York University Langone Medical Centre was

Indian Healthcare's Climate Footprint

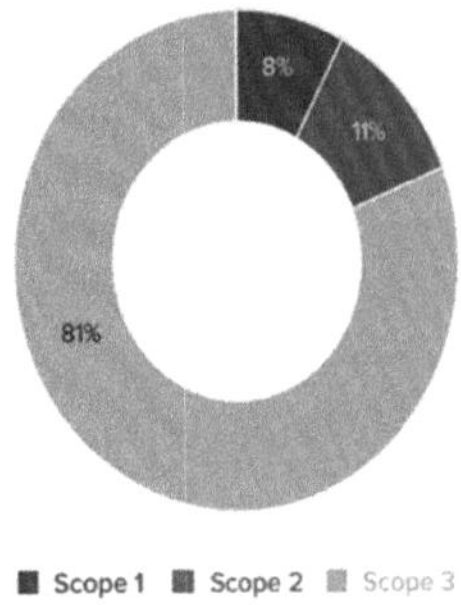

India

India health care	Value	Unit
Climate footprint	39	$MtCO_2e$
Emissions per capita	0.03	tCO_2e/ capita
Emissions as % of national footprint	1.5	%
Expenditure per capita	57	USD
Expenditure as percentage of GDP	3.6	%
% of footprint generated domestically	80.1	%
Health sector footprint equivalence to coal power plant emissions[32]	10	coal-fired power plants in one year

SCOPE 1 - Take immediate action to reduce healthcare facility emissions.

SCOPE 2 - Invest in and advocate for the decarbonization of local and national energy systems and the implementation of clean, renewable energy.

SCOPE 3 - Set and implement criteria for low-carbon or zero-emissions procurement so as to begin to decarbonize the supply chain.

Source: Healthcare's Climate footprint September 2019

Fig. 6.1 India's healthcare climate footprint. (Adapted from Karliner et al. 2020, Health Care Without Harm. Arup 2019)

one the most damaged structures (World Health Organization 2020). It was fully closed for around 2 months and without an emergency room for a year and a half. There are several similar examples, and these are just two of them. Thus, serious efforts are required to combat the effects of climate change and to minimize their impact on healthcare infrastructure and the overall health of the communities.

6.3 Climate-Smart Approach to Building Health Systems Resilience

The transformative action towards climate-smart healthcare can be in two dimensions, namely, adaptation or resilience and mitigation and greening. There is a need to have a climate-prepared and decarbonized health system. The Malé Declaration on Building Health System Resilience to Climate Change, 2017, saw all Member States, including India at the 70th session of the WHO Regional Committee for South-East Asia committed to building health system resilience to climate change (Regional Committee for South-East Asia 2017). The commitments under this Declaration are fundamentally linked to the seven key elements of climate-smart healthcare (International Bank for Reconstruction and Development & The World Bank 2017):

- Overall system design for coordinated care with emphasis on local providers and driven by public health needs
- Building design and construction based on low carbon approaches
- Investment in renewable energy and energy efficiency

- Waste minimization and sustainable healthcare waste management
- Sustainable transport and water consumption policies
- Low carbon procurement policies for pharmaceuticals, medical devices, food and other products
- Resilience strategies to withstand extreme weather events and anchor sustainable communities

A fundamental aspect of providing safe and quality services within climate-resilient and environmentally sustainable healthcare facilities is the health workforce—not only in numbers but also in skills; human resources should be empowered and informed to respond to environmental challenges (World Health Organization 2020). The next area of focus is ensuring water, sanitation, hygiene, and healthcare waste management are sustainable and safe. Other important requirements are safe and sustainable energy and related services, infrastructure, technologies, products, and processes that include appropriate technology, products, and processes that allow efficient functioning of the facilities.

6.4 Evolution of India's Policy Response to Climate Change

In 2006, the National Environment Policy was formulated, and the Ministry of Environment, Forest and Climate Change (MoEFCC) was nominated as the nodal Ministry. Prime Minister's Council on Climate Change was formed in 2007 to advise proactive measures to facilitate inter-ministerial coordination and guide policy. The National Action Plan on Climate Change (NAPCC) was released by the then Honourable PM in June 2008 (Press Information Bureau 2021). India has been taking climate change quite seriously and eight national missions were launched in 2008:

- National Mission on Sustainable Habitat
- National Mission for Sustaining the Himalayan Ecosystem
- National Mission for Sustainable Agriculture
- National Solar Mission
- National Mission for Enhanced Energy Efficiency
- National Water Mission
- National Mission on Strategic Knowledge for Climate Change
- National Mission for Green India (DST & GoI 2025)

The Expert Committee assessed the impact of climate change on six areas: Natural Ecosystem, Health, Agriculture, Water Resources, Coastal Zone Management, and Climate modelling. Reports of the Expert Committee highlighted a range of policies and programs to address the problem of climate change in the context of sustainable development.

The NAPCC outlined a national strategy to enable climate change adaptation and enhance the ecological sustainability of India's development path. It emphasized that to increase the living standards and quality of life of the people of India, maintaining a high growth rate is essential, and at the same time, reducing their vulnerability to the impacts of climate change is also indispensable.

In 2015, progress was reviewed, the Prime Minister Council on Climate Change was reconstituted, and four new missions were added to the already existing basket of eight missions (Press Information Bureau 2014). Health being the foremost priority, Health Mission has been added to these four new additions. The other missions were the National Mission on 'Waste to Energy Generation', the National Mission on India's Coastal Areas, and the National Wind Mission. Hence, out of 12 missions, health is at the centre in the revised scheme of things. For the Health Mission under NAPCC, MoHFW is the nodal agency that constitutes a national expert group on climate change and health. They were tasked to prepare an action plan, recommend strategies for adaptation and response, and plan for diseases occurring due to climate variability and change. The National Centre for Diseases Control (NCDC) is the nodal designated by MoHFW for all these activities (MoHFW & GoI 2018).

6.5 National Program for Climate Change and Human Health (NPCCHH)

NPCCHH launched in 2019, envisaged the vision of strengthening the healthcare services for all the citizens of India, especially vulnerable populations, children, women and marginalized against climate-sensitive diseases (CSDs). The goal was to reduce morbidity, mortality, injuries, and health vulnerability due to climate variability and extreme weather (NCDC 2023; MoHFW 2018). The objectives were to strengthen healthcare services against the adverse impact of climate change on health. The specific objectives are to:

1. Create awareness among the general population, especially vulnerable populations; healthcare providers, and policymakers regarding the impact of climate change on human health.
2. Strengthen the capacity of the healthcare system to reduce illnesses due to variability in climate.
3. Strengthen health preparedness and response by performing situation analysis at all levels.
4. Develop partnerships and create synchrony and synergy with other missions and ensure that health is adequately represented in the climate change agenda in the country.
5. Strengthen research capacity to fill the evidence gap on climate change's impact on human health (MoHFW & GoI 2018).

The overall concept is to establish healthcare facilities that are environmentally sustainable (mitigation) and climate-resilient (adaptation). The focal point in green healthcare facilities is their responsibility to adopt sustainable carbon solutions; climate-resilient healthcare infrastructure is designed to adapt, reorganize, and evolve to be prepared for better future preparedness for disasters or public health emergencies (MoHFW & GoI 2018).

The strategic framework under NPCCHH focuses on strengthening and coordinating health-related early warning and surveillance activities through an integrated disease surveillance system. Under the Pradhan Mantri Ayushman Bharat-Health Infrastructure Mission (PM-ABHIM), the strategy is to strengthen/develop coordination for the disease surveillance system in place, covering not just the physical infrastructure but also leveraging the technology of the Integrated Health Information Platform (IHIP).

A second component is the feedback mechanisms with other ministries overseeing the ecological determinants that influence health including, air quality, water, food supply, energy sources, and human resources. MoHFW is working closely with all other ministries on aspects related to climate change. Risk mapping for each geographical area is being done by the National Centre for Disease Control (NCDC). Additionally, four new branches of NCDC are coming up very shortly under PM-ABHIM and their linkages with State Public Health Labs, District Public Health Labs (IPHLs), and Block Public Health Units (BPHUs) are also being streamlined.

Innovative or new strategies for technical approaches, increasing access to early healthcare, referent health tracking, and incorporating unique IDs are all additional strategies to bolster health system strengthening and resilience. For example, the Ayushman Bharat Digital Mission (ABDM) has been launched with this in mind. Under this mission, digital health IDs are being created such that every individual's health record will be available in a longitudinal manner. All the applications currently being used in the health system will be part of the ABDM stack and based on this, data will be available with dashboards at all levels starting from block onwards and even with the health functionaries. This arrangement will help in coming up with newer innovative approaches for accessibility of healthcare services, early advice/consultation, early referral, and early response by the health system in times of emergencies. Conducting case studies, research, and pilot testing of new approaches is essential to build health resilience in climatically sensitive regions.

6.6 Framework Under NPCCHH for Green Healthcare Facilities

According to the Malé Declaration, healthcare facilities should be designed in a way to ensure uninterrupted operation of essential services such as water supply, sanitation, waste management, and electricity remain functional during climate-related events. In addition, the health department must implement measures to make the health sector more sustainable by adopting eco-friendly technologies and energy-efficient services. NAPCCHH recognizes and reinforces all these recommendations (MoHFW 2018).

To support the NPCCHH's objectives, the concept of green hospitals has been integrated into public healthcare facilities through the existing Kayakalp scheme.

As part of this initiative, a new theme of 'Eco-friendly Facilities' has been added, ensuring that facilities/hospitals remain clean, green, infection, and pollution-free and also fosters faster healing. Additional key themes include enhancing energy efficiency; minimizing air and noise pollution; reducing, reusing, and recycling of waste; saving the earth and environment; and supporting overall health and well-being (NHSRC 2023a).

For the development of green healthcare facilities or infrastructure, guidelines have been issued for Brownfield or Greenfield land projects in India (NCDC 2023). Guidelines for building and infrastructure have been developed, including what kind of land to be used, and natural resources such as air, water, energy, and food. For energy sources, the on-grid solar as well as off-grid models are being considered. NHSRC recommends that in facilities such as District Hospital and above where the grid is easily available, arrangements for on-grid could be made with private players through public-private partnerships. For sub-district health facilities, which are far off from the grid, an off-grid model has been recommended. Issues like maintenance of batteries in the off-grid model, changing batteries every 7 years, and increasing battery life will then need to be addressed. Other aspects include waste management including sewage treatment plants and effluent treatment plants (ETPs) for every facility. Sustainable procurement mechanisms are being established through utilizing the 'Make in India' campaign. The biomedical equipment management program is now being strengthened. Going beyond procurement, reducing the wastage of equipment, and increasing the life of equipment through preventive maintenance, and regular repairs, are integral components within this.

Procurement systems are being strengthened across the country. Under the transport, communications, and information services, the Ministry of Road, Transport and Highways has already issued guidelines for ambulances. Discussions are being conducted with them on how Ambulances or Patient Transport Vehicles (PTVs) could be made more eco-friendly. Site selection of facilities plays an important role in preventing adverse effects on the climate. The site selection for healthcare facilities may also alter the impact of disasters like flooding or earthquakes in the area.

Therefore, to achieve climate-resilient and green healthcare facilities, the mapping of climate vulnerability and disease vulnerability through granular statistics, including local healthcare utilization patterns, is being prioritized. The Vision for 2025–26 outlines the current measures taken. Health audits (personnel, stocks, and infrastructure), energy audits (for lighting, heating, cooling, and equipment), capacity-building training workshops, webinars, and guidance resources are being done by the NCDC and other nodal agencies. Communication materials on air pollution, heatwaves, VBDs, water-borne diseases, post-traumatic stress disorders, etc., have been prepared and are beginning to be disseminated. Adequate financial resources have also been allocated for these initiatives under respective programmatic umbrellas. Awareness generation and advocacy to involve communities and local leaders are ongoing. Currently, 42 health days are being observed at every Ayushman Bharat-Health and Wellness Centre (AB-HWCs) in the country, and the environment forms an integral part of these (NHSRC 2023b).

Discussions are ongoing with the manufacturers to look at energy-efficient medical equipment including lighting, cool roofs, rainwater harvesting, wastewater recycling with good water metering systems (to prevent wastage of water), safe and appropriate waste handling, and different types of sewage and effluent treatment plants. All these are being made mandatory for the newer facilities. For older facilities, retrofitting and installation of rainwater harvesting systems, fitting of solar energy plants where the rooftops are available, use of LED lighting (whenever lights are to be changed, they are to be replaced by LEDs), etc., are being encouraged. For non-biomedical wastes, effluent treatment plants (ETPs) are being made as part of all new constructions, and facilities' connection with the local sewage treatment plant is being emphasized. This is an incremental approach, working closely across stakeholders at national and sub-national levels.

Environment Health Cells have been established in various State (provincial) Health Departments. State Nodal Officers (SNOs) have been identified for climate change in State Health Departments and State Department Task Forces are being formed with representation from various health programs, departments, and ministries. State-specific health adaptation plans are being prepared for extreme weather events, following the local context. The health departments have been asked to identify and build capacities for monitoring and surveillance of climate-sensitive diseases (CSDs). Premier institutes are being involved in the development of training modules and guidelines for climate change for the health workforce. These institutes are also being utilized for the implementation of a media communication plan for CSDs, including health and health-determining sectors and communities.

In the coming 2–5 years, the next phase will encompass the development of a specific implementation framework for CSDs including capacity building and training of healthcare personnel on guidelines and treatment modalities, tools for diagnostics, surveillance, and mathematical or prediction models for CSDs. The plan will also include periodic reviews for monitoring of indicators (vulnerability response planning and capacity) to develop preparedness against environmental determinants for each CSD.

Adaptation of newer technologies, building designs, and efforts related to energy, water, and sanitation—will be included in the action plan. Efforts are being undertaken to come up with newer kinds of building designs that are high energy saving and with water and sanitation provisions—which have been made mandatory now. Linking data on CSDs, environmental factors, and outcomes such as morbidity and mortality will improve, as PM-ABHIM is implemented over the next 2–4 years. Risk mapping and seasonal trends for CSDs including a multi-sector management approach, along with research and epidemiological studies/surveys on the vulnerable population of CSDs, are also being initiated.

The expected outputs from the program is to increase public awareness and encourage behaviour changes concerning the impact of climate change on health and draws attention to prevention and proactive measures towards mitigation. Furthermore, it looks at expanding the number of trained health personnel and equipped institutions, and improving the capacity of organizations towards delivering these essential services and infrastructure across the country. It also looks at an integrated monitoring system for the collection and analysis of health-related data, along with metrological, environmental, socio-economic, and occupational factors.

Continuous monitoring and evaluation of each of these activities, development of favourable legislations and policies, training for healthcare staff to the village level, collation and dissemination of the good practices, and commissioning of impact studies are all additional efforts that are required.

6.7 Sub-national Focus Areas

Each state is different from the other with each having its own set of challenges and opportunities. In general, southern states, like Tamil Nadu, are much ahead in terms of the provision of healthcare services and focus on climate change aspects as well such as the usage of solar and other renewable sources of energy. State ranking has not been done yet under the program because all states are in the process of moving towards climate-resilient facilities as well as implementing GoI policies. Overall, each state is putting in tremendous effort to achieve the desired outcomes while addressing the challenges and learning from the best practices of other states/UTs. North-eastern states and Odisha are working on the provision of solar energy for their facilities, especially in the newer facilities being established.

Activities envisaged at the district level under the NPCCHH program have three major components, namely, awareness generation, capacity building, and surveillance. Awareness generation consists of IEC/BCC campaigns, public health advisories, observance of important environment and health days, along monitoring and supervision of these activities. Under the capacity building component, training on climate change and health for healthcare workers and communities is envisaged. Community engagement platforms associated with primary and secondary care facilities are being utilized for sensitization on climate change and health. Surveillance for heat-related illnesses as well as air pollution-related illnesses is also being done in all states.

Flexibility has been provided to States/UTs to develop their action plans in keeping with the local context, in terms of geography, climate conditions, disaster situations, urbanization, and available resources such as finances, and health infrastructure. Each state is now coming up with its initiative to prevent and/or mitigate the effects of climate change. Some examples can be seen such as the dental unit of the primary health centre in Gumbhalli, Karnataka, or the cold chain in Manipur which are running on solar energy, portable solar-powered COVID-19 testing kiosks used in Meghalaya (Centre for Environmental Health & PHFI, n.d.). Efforts are being made to run basic equipment and facilities with solar energy in model health centres and mobile clinics, especially for remote geographies. Solar power is also being considered to improve safety and comfort such as in maternity waiting homes. In some instances, larger healthcare facilities are also being powered by solar energy such as the 100-bed in-patient unit and the 50-patient chemotherapy unit in Bihar built by the NGO 'Doctors for You'. Several district hospitals are looking at the on-grid model for energy requirements as well.

6.8 Other Government Health Programs with Impacts on the Environment

Decarbonizing healthcare is the need of the day. The four pillars of Ayushman Bharat, two (AB-HWCs and AB-PMJAY) launched in 2018 and the other two (PM-ABHIM and ABDM) launched in 2021, aim to create a strengthened and resilient health system. The provision of comprehensive primary healthcare near the home of people with a time-to-care approach of 30 minutes and assured linkages with secondary and tertiary services and a continuum of care pathways will save people's transport requirements to the healthcare facilities. This is another part of the decarbonization of healthcare (NHSRC 2018; MoHFW 2021a, b). Digital health interventions will further save travel, opportunity cost, daily wages, and requirement of fuel and vehicles.

Environment-friendly features are being supported by the GoI at public health facilities. IPHS, 2022 (NHSRC 2022) mandates 60 percent of the total area of the facility to be reserved for landscaping, play area, and circulation area and the remaining 35–40 percent should be the plinth area that includes the residential area too. This will reduce the travel of the healthcare workers to the health facility every day. Vertical expansion is recommended instead of horizontal to ensure more space for creating green spaces and in areas with space constraints (especially in urban areas).

Similarly, rainwater harvesting, biomedical waste management, water-saving equipment for sterilization and mechanical laundry are being strengthened.

PM-ABHIM is the largest infrastructure mission and scheme in our country. The central sector component of PM-ABHIM includes strengthening surveillance of infectious diseases and outbreak response across the country through the creation of Integrated District Public Health Labs (IPHL) and Block Public Health Units (BPHU). It is anticipated that climate change will unleash public health emergencies at regular intervals and only resilient health systems can prevent it from becoming an emergency and manage these by identifying early warning signals and establishing linkages and capabilities. IPHLs have a provision of Epidemiologists and Entomologist for better surveillance in the area; hence, linkages will be strengthened along with the capacity to undertake various tasks. Disaster and epidemic preparedness is being strengthened through the creation of 15 new health emergency operation centres and two container-based mobile hospitals. Further, a national institution for One Health, a regional research platform for the WHO Southeast Asia Region, and advanced bio-safety laboratories are being set up.

6.9 Collaboration and Intersectoral Coordination

Intersectoral coordination is the key to success for all government initiatives on climate change. At the national level, many efforts are being made to have better coordination with different ministries and departments for a holistic impact. Intersectoral coordination is gaining pace amongst Ministries of Jal Shakti (Water

Affairs); Environment, Forest and Climate Change; Housing and Urban Affairs; Panchayati Raj; and Health and Family Welfare.

There is a definitive opportunity to strike effective and congenial collaboration with academicians, government, NGOs, and civil society organizations to build resilient health systems. All the government initiatives at various levels will have an impact on India's ability to steer climate and health security concerns on the global stage, including through the current Indian Presidency of the G20 forum. The Indian government is also drawing lessons from COVID-19 for designing future initiatives.

The developed world must also do much more than what they are currently doing. India is responsible for only 6–7 percent of the emissions that are contributing to global climate change. The world should understand India's developmental concerns and link the climate change targets accordingly. The Health Ministry will showcase how India is addressing climate change from a health perspective through best practices, District IPHLs, and BPHUs, surveillance systems, etc.

National Health Systems Resource Centre (NHSRC) is committed to being a part of the solution with all divisions of NHSRC being part of this initiative. The Quality and Patient Safety (QPS) division is the nodal for National Quality Assurance Standards (NQAS) and Kayakalp scheme, and it has issued guidelines for biomedical waste management. The Public Health Administration (PHA) division has come up with the IPHS standards 2022, and it was involved in the framing of NPCCHH guideline in 2019. Community Processes and Comprehensive Primary Health Care (CP-CPHC) Division developed Jan Arogya Samiti (JAS) and Wellness guidelines and their linkages with climate change (NHSRC 2023d). The Human Resources for Health and Health Integrated Planning (HRH/HPIP) division is responsible for capacity building of healthcare workers. Knowledge Management Division (KMD) takes care of the implementation research, evidence generation, and knowledge management related to health system strengthening. The Healthcare Technology (HCT) division looks after the technical specifications and technology evaluation and how to align them with the climate change policies and programs. Healthcare Financing (HCF) is involved in the National Health Accounts that provide information about the financing aspects of the health system, including sources of funds, fund allocation, utilization, analysis and documentation. The IT division is working towards an effective digital ecosystem (Fig. 6.2).

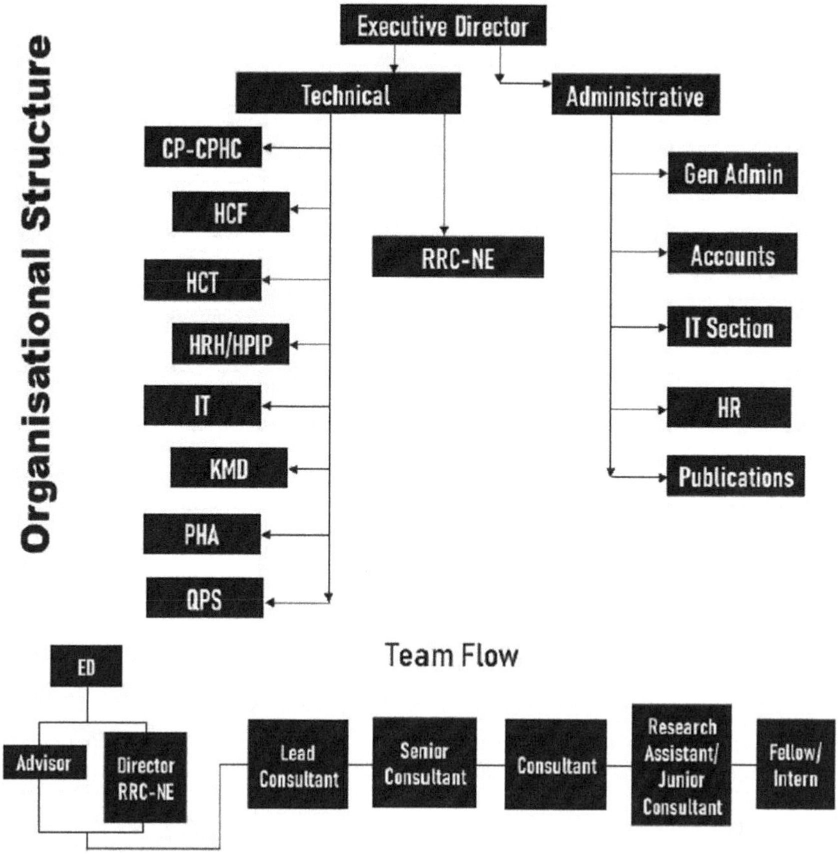

Fig. 6.2 NHSRC structure. (Adapted from NHSRC India)

6.10 Conclusion

Climate change is one of the most important challenges of this century, which may have an overwhelming impact and consequences if not addressed with urgency, efficiency, and effectiveness. Realizing the need for a robust action plan for climate change and building resilient health systems, the government of India has taken several initiatives under the umbrella of the NPCCHH and linked them with several other schemes and programs as elaborated in this chapter. These plans and programs need to be systematically and efficiently implemented on the ground. Additionally, there is a need felt to develop indices for measuring progress on climate change adaptation, beyond those currently outlined in the NPCCHH. The government needs to develop impact evaluation strategies and formulate specific indicators for climate change. Universities and academicians can be involved in developing these scientifically robust strategies and indices.

As climate change is interdisciplinary and transnational, inter-ministerial and intersectoral coordination is integral. The 12 missions under the NAPCC should interact with each other and integrate their activities. Partnerships with academic institutions, the private sector, NGOs, and CSOs can be systematically engaged in line with the public health goals. There is also scope for transnational collaborations to exchange evidence, knowledge, and best practices. All these measures, if implemented in a timely fashion, will go a long way towards climate resilience and enhancing the quality of life of people.

References

Centers for Disease Control and Prevention (CDC). CDC's Climate and Health Program, 2022. https://www.cdc.gov/climateandhealth/default.htm.

Centre for Environmental Health, PHFI. Green and Climate resilient healthcare facilities, n.d.. https://ncdc.gov.in/WriteReadData/linkimages/GreenClimateResilienthealthcarePHFI.pdf.

DST, & GoI. Climate change Programme | Department Of Science & Technology. Department of Science & Technology, 2025, April 28. https://dst.gov.in/climate-change-programme.

Health Care Without Harm. Health care climate footprint report, 2019. https://noharm-uscanada.org/climatefootprintreport.

Karliner J, Slotterback S, Boyd R, Ashby B, Steele K, Wang J. Health care's climate footprint: the health sector contribution and opportunities for action. Eur J Pub Health. 2020;30(5) https://doi.org/10.1093/eurpub/ckaa165.843.

Ministry of Environment, Forest and Climate Change, Government of India. India's long-term low-carbon development strategy, 2022. https://moef.gov.in/wp-content/uploads/2022/11/Indias-LT-LEDS-2.pdf.

Ministry of Health and Family Welfare, Government of India. National Action Plan for climate change & human health, 2018. https://ncdc.gov.in/WriteReadData/l892s/27505481411548674558.pdf.

Ministry of Health and Family Welfare, Government of India. Operational guidelines for PM Ayushman Bharat Health Infrastructure Mission, 2021a. https://nhsrcindia.org/sites/default/files/OGls%20PM%20Ayushman%20Bharat%20Health%20Infrastructure%20Mission.pdf.

Ministry of Health and Family Welfare, Government of India. Implementation of 15th Finance Commission (FC-XV) - Health Grants Through Local Governments, 2021b. https://nhsrcindia.org/sites/default/files/2021-09/FCXV%20Technical%20and%20Operational%20GLs%20to%20States%20dated%2031082021.pdf.

MoHFW, & GoI. National Action Plan for climate change and human health. Ministry of Health and Family Welfare; 2018. p. 1–70. https://ncdc.mohfw.gov.in/wp-content/uploads/2024/04/27505481411548674558.pdf

National Centre for Disease Control. National Programme on Climate Change & Human Health (NPCCHH), 2023. https://ncdc.mohfw.gov.in/index1.php?lang=1&level=1&sublinkid=876&lid=660.

National Health System Resource Centre. Ayushman Bharat: comprehensive primary health care through health and wellness centers - operational guidelines, 2018. https://nhsrcindia.org/sites/default/files/2021-03/Operational%20Guidelines%20For%20Comprehensive%20Primary%20Health%20Care%20through%20Health%20and%20Wellness%20Centers.pdf.

National Health System Resource Centre. IPHS 2022 and its layouts, 2022. https://nhsrcindia.org/IPHS2022.

National Health System Resource Centre. Kayakalp and Swachh Swasth Sarvatra, 2023a. https://qps.nhsrcindia.org/kayakalp-swachh-swasth-sarvatra.

National Health System Resource Centre. Comprehensive primary health care, 2023b. https://nhsrcindia.org/practice-areas/cpc-phc/comprehensive-primary-health-care.

National Health System Resource Centre. Emergency COVID Response Package (ECRP), 2023c. https://nhsrcindia.org/ecrp-ii.
National Health System Resource Centre. About us and NHSRC's Work Areas, 2023d. https://nhsrcindia.org/.
Patz JA, Campbell-Lendrum D, Holloway T, Foley JA. Impact of regional climate change on human health. Nature. 2005;438(7066):310–7. https://doi.org/10.1038/nature04188.
Press Information Bureau. Government of India, Ministry of Environment, Forest and Climate Change, Government reconstitutes the Prime Minister's Council on Climate Change, 2014. https://pib.gov.in/newsite/printrelease.aspx?relid=111090.
Press Information Bureau. Ministry of Environment, Forest and Climate Change: National Action Plan on Climate Change (NAPCC), 2021. https://static.pib.gov.in/WriteReadData/specificdocs/documents/2021/dec/doc202112101.pdf.
Regional Committee for South-East Asia, 70. Male declaration on building health systems resilience to climate change. World Health Organization. Regional Office for South-East Asia, 2017. https://apps.who.int/iris/handle/10665/259578.
World Bank Group. Climate-Smart Healthcare. Low-carbon and resilience strategies for the health sector, 2017. https://documents1.worldbank.org/curated/en/322251495434571418/pdf/113572-WP-PUBLIC-FINAL-WBG-Climate-smart-Healthcare-002.pdf.
World Health Organization. WHO guidance for climate resilient and environmentally sustainable healthcare facilities, 2020. https://www.who.int/publications/i/item/9789240012226; https://www.who.int/news/item/12-10-2020-who-publishes-guidance-on-climate-resilient-and-environmentally-sustainable-health-care-facilities.

Air Pollution and Climate Change: A Public Health Challenge in India

7

Sarang Satheesh and Satheesh Kumar

Abstract

Air pollution and climate change are two of the most vital environmental threats of this century, as their effects are being severely felt around the world. Floods, fires, heat waves, large-scale melting of polar ice caps, rising sea levels, and changes in wind and ocean circulation are direct consequences of global warming. In India, the occurrence of festivals has been linked to increased air pollution and associated health risks. The coordinated use of pyrotechnics increases air pollution in the case of global festival occurrences. This chapter analyses the emission of toxic air pollutants from fireworks and the resulting poor air quality during the Vishu festival in Kannur, Kerala, as a case study to illustrate health issues. This has been linked with a sudden increase in respiratory cases among different age groups, demonstrating that children and the elderly are the most vulnerable to the increasing air pollution caused by fireworks.

Keywords

Climate change · Air pollution · Fireworks · Vishu · Festivals · India · Kannur

7.1 Introduction

Air pollution and climate change are challenging environmental issues that have been the subjects of pioneering scientific research for more than a century. The studies on air pollution focus on its potential contribution to global warming as well as

S. Satheesh
General Electric, London, UK

S. Kumar (✉)
Manipal Academy of Higher Education, Manipal, India

H. Brand et al. (eds.), *Climate Change and Public Health*,
https://doi.org/10.1007/978-981-95-0895-2_7

its impact on human health and ecosystems. The warming of the atmosphere leads to climate change, which can lead to extreme weather conditions that can harm public health, both directly and indirectly, as outlined by other authors in this book.

Uncontrolled and massive burning of fossil fuels produces GHGs that warm the planet, and thermal energy is released into the atmosphere, heating the atmosphere and the earth (IPCC 2018). This dumping of energy is classified as radiative forcing (RF). The main RF-positive GHGs are carbon dioxide (CO_2), water vapour (H_2O), methane (CH_4), nitrous oxide (N_2O), halocarbons (HC), sulphur hexafluoride (SF_6), and ozone (O_3), which are produced by industrial and anthropogenic activities (Crutzen et al. 1979). A detailed study (Bellouin et al. 2020) after performing reanalysis modelling revealed the mean global RF values for the atmospheric pollutants (Table 7.1).

The values in brackets show a statistically significant variation of 95 percent. This demonstrates the positive levels of RF of CO_2, CH_4, and surface O_3 warming the planet, as well as the cooling effect of stratospheric ozone and aerosols. Santer et al. conducted a fingerprint analysis using satellite-derived patterns of temperature variation that span from the lower troposphere to the upper stratosphere to estimate the impact of the stratosphere on climate change (Santer et al. 2023). As the efficiency and global warming potential of each gas differ, the rate of warming varies with the mixing ratio of each gas in location, which greatly influences the chemistry and transport of pollutants in the atmosphere (Solomon et al. 2009).

The fine particulate matter (PM) that pollutes the air contains mainly sulphates, nitrates, soot, and secondary organic aerosols (SOA). These particles scatter sunlight, leading to a reduction in sunlight on the Earth's surface, and their RF values are negative due to the cooling of the atmosphere (Brasseur et al. 2006). At the same time, the soot particles released during combustion absorb sunlight to generate heat, becoming local heat centres and considered significant pollutants that can further modulate climate.

Since the scattering of sunlight by particles leads to cooling while GHGs warm the surface, there is uncertainty about the predominance of particles or GHGs. This adds complexity to the scientific analysis of air pollution and climate change, which are essentially two sides of the same coin.

Table 7.1 Comparison of best estimate and 5–95% confidence ranges for stratospherically adjusted radiative forcing (in Wm^{-2})

Radiative-forcing agent	IPCC AR5 estimate	Study by Bellouin et al. in 2020
CO_2	+1.82 (1.63 to 2.01)	+1.91 (1.51 to 2.31)
CH_4	+0.48 (0.43 to 0.53)	+0.46 (0.36 to 0.56)
Tropospheric ozone	+0.40 (0.20 to 0.60)	+0.33 (0.01 to 0.59)
Stratospheric ozone	−0.05 (−0.15 to 0.05)	0.00 (−0.20 to 0.20)
Aerosol–radiation interactions	−0.35 (−0.85 to +0.15)	−0.56 (−0.90 to −0.23)
Aerosol–cloud interactions	−0.45 (−1.2 to 0.0)	−0.71 (−1.14 to −0.44)

By Bellouin et al. (2020)

7.2 Air Pollutants

GHGs such as carbon dioxide (CO_2), nitrogen oxides (N_2O), water vapour (H_2O), methane (CH_4), and surface ozone (O_3) have higher global warming potential, and their increasing concentrations could lead to climate change (Abbass et al. 2022). Increased human activities, industrial transport, and vehicles lead to emissions of CO_2 and N_2O, while CH_4 is mainly produced from wetlands and paddy fields as soil microbes and plants are forced to anaerobically metabolize into methane, which has a 25 times greater greenhousc effect than CO_2 (Zhang et al. 2016).

The main source of N_2O comes from various soil management activities in agriculture, such as the widespread use of organic fertilizers to increase yields and the burning of agricultural residues. Ground-level ozone is a secondary air pollutant that forms from its main precursors, such as NO_2, CO, and CH_4, in the presence of sunlight. O_3 is a strong oxidizer in the atmosphere and has a high heat storage capacity, which means that O_3 has about 1000 times higher heating efficiency than CO_2. Smog is created by ozone, which is abundant in all major cities during winter, making O_3 an ideal indicator for estimating air pollution in the region (Resmi et al. 2021; Lal et al. 2000; Wang et al. 2022).

Air pollutants are transported over long distances by atmospheric winds. For example, it takes a week for air pollutants to travel from East Asia to the west coast of North America and across the Atlantic to Europe after crossing the Pacific Ocean. Because the lifespan of air pollutants typically ranges from a few days to several weeks, depending on the environment, air pollutants can spread across the planet in weeks. Air pollutants are typically removed from the air by wet and dry deposition on surfaces and by rain. Therefore, GHGs, except for O_3, emitted in every corner of the planet, can spread to other places due to their long duration (more than a decade), store heat, and affect the global climate. For this reason, this phenomenon is called global warming. These scientific facts are very important in the fight against air pollution, which often requires local, regional, national, and even global action.

The COVID-19 lockdown led to an unprecedented decline in global economic activity and associated air pollution, largely due to reduced land transport, and this has been widely reported around the world (Saha et al. 2022). India witnessed a significant decline of 43, 31, 10, and 18 percent in $PM_{2.5}$, PM_{10}, CO, and NO_2 concentrations, respectively, during the lockdown period when compared to previous years (Shubham et al. 2020). The significant reduction in air pollution and rapid improvement in air quality shed light on a long-term emergency management strategy for efficient mitigation.

7.3 Climate Change

The term 'climate' refers to the average conditions of the weather at a given point on the Earth's surface. Generally, climate is measured in terms of temperature, rainfall, and wind patterns based on past data. "Climate change' is a shift in the average climate or the variability of the climate over an extended period.

The cause of climate change is the shift in the Earth's energy balance. This means that more of the sun's energy reaches the Earth and its atmosphere but is then reflected into the universe. Consequently, additional emissions of GHGs discharged into the atmosphere, forming a 'blanket' that ensnares warmth from the Earth's surface, thereby leading to a rise in temperature. The warming of the atmosphere and the ground leads to a change in the energy balance over time, which leads to a change in the atmospheric and ocean circulation pattern (Topál and Ding 2023).

Air pollution can thus contribute to increasing frequency/intensity of heat waves. Urban sprawl further increases the number of built-up areas that produce heat, increases the sealing of areas, and necessitates longer travel distances. Thus, heat waves consequently result in the formation of urban heat islands (UHI) that differ from city to city depending on changes in land use. It is uncovered that a rising pattern happened mostly due to growth in built-up areas and a decrease in vegetation (Jabbar et al. 2023). A numerical analysis is documented on the influence of the land surface temperatures (LST) on the magnitude of the surface urban heat island (SUHI) across 42 cities in India and 32 cities in the United States using satellite imagery to obtain their impact on climate change (Tetali et al. 2022). Extremely hot climates due to increased air pollution near the Mediterranean Sea and in West and Central Africa are also more prone to extreme droughts (Cramer et al. 2018). Similarly, bushfires/forest fires resulting from excessive heat can in turn lead to increased air pollution. For example, recent reports of fierce bushfires in Australia and Canada in 2019 and 2023, respectively, were fuelled by high temperatures.

Elevated air temperatures are expediting the thawing of the extensive glaciers and ice sheets of Antarctica and Greenland, which can cause flooding. The average global sea level rise has been estimated at 4.5 mm per year between 2013 and 2021 (Durand et al. 2022). Surprisingly, sea level rise in the Indian Ocean region is the fastest in the world due to sea surface temperature (SST) anomalies in the Indian Ocean. According to the Government of India's Department of Geosciences, the mean sea level along the Indian coast rose by 1.7 mm per year over the last century between 1900 and 2000 (Swapna et al. 2020). This has increased the likelihood that storm surges—the rise in sea level during severe storms—will breach existing sea barriers and cause flooding. Many of the densely populated cities are in low-lying coastal areas around the world, sea level rise poses a major threat that could affect many people, especially in economically developing countries.

Evaporation from the sea increases with rising temperatures and warmer air can hold more water, resulting in more precipitation (Blenkinsop et al. 2021). The Indian monsoon pattern has experienced noticeable variations since 1951 (Goswami et al. 2022). It has decreased, specifically in the western and Indian-Gangetic plains (IGP) areas, while the regionally intense precipitation in central Indian states has amplified. There has also been a substantial increase in the number of dry spells in the intervals between rainy days during the Indian monsoon. The Indian floods in 2018 and 2019 caused disaster-like events such as landslides, river floods, and flash floods in one of the world's most disaster-prone regions, leading to deaths and severe destruction. At the time of writing this, several regions in northwest India are encountering intense flooding and torrential downpours, with the Indian capital Delhi surpassing a rainfall record that had stood for 41 years.

7.4 Health Impacts of Air Pollution

The effects of air pollution and climate change on human life are far-reaching, as both have been identified as potential health risks in the present century (Brunekreef and Holgate 2002; Noël et al. 2021; Ramanathan 2020). India, the world's most populated nation, is particularly vulnerable to air pollution, both indoors and outdoors, which has a considerable effect on health and economic development. While the health effects of air pollution are most pronounced in urban areas, mortality rates are also significantly high in rural areas. This is because most of the people living in rural regions depend on the utilization of biomass or kerosene for cooking and lighting, which deteriorates indoor air quality and can result in severe health issues. India's rapidly expanding megacities and the increasing number of vehicles on the roads have exacerbated air pollution (Gurjar et al. 2016; Molina and Molina 2004). The air quality in north India is especially poor due to the huge amount of stubble burning and trapping of pollutants in the cold Himalayan valleys (Saikawa et al. 2019; Shukla 2021).

Poor air quality contributes to the prevalence of lung diseases, such as asthma, bronchitis, lung cancer, chronic obstructive pulmonary disease (COPD), birth defects, immune system disorders, cardiovascular problems, and pneumonia (Mc Evoy and Buggy 2023). According to a study published in *The Lancet Planetary Health*, air pollution caused about 1.6 million deaths in India in 2019, the highest number in the world (Fuller et al. 2022). Most deaths in India were caused by particulate matter of size 2.5 μm (PM2.5) by domestic biomass burning, followed by coal and crop burning. Deaths from automotive and industrial air pollution have also reportedly increased by 55 percent since 2000.

Another important study by David et al. (2019) used modelling to predict 1.1 million premature deaths per year from $PM_{2.5}$ pollution in India alone. The causes of premature mortality constituted about 60 percent of local anthropogenic pollution in India, ~19 percent was due to the transport of pollutants between different parts of the country, about 16 percent was due to long-distance transport of anthropogenic pollutants outside of India, and about 4 percent was related to natural $PM_{2.5}$ sources. Almost 46 percent of India's premature mortality rate is found in the Indo-Gangetic plains due to vulnerability to air pollution, with southern India accounting for about 13 percent. Almost eight percent of premature deaths were caused by the transport of air pollutants from the Indo-Gangetic plains to places with higher premature death rates in India. The model also estimates that the hotspots with more than 50 percent premature mortality due to the transport of $PM_{2.5}$ from the peripheral regions are in the northern, eastern, western, and central parts of India.

7.4.1 Acute Air Pollution and Associated Health Risk

Some catastrophic occurrences, such as forest fires, fireworks used during festivals, and volcanic eruptions, can swiftly pollute a place. The abrupt increase in occurrences has detrimental health implications, and these incidents might contribute to local pollution. Rock concerts, major sporting events' opening and closing ceremonies, amusement parks, and political wins all frequently feature fireworks displays. In the United States, for example, amusement parks are the greatest consumers of fireworks, and the Fourth of July has the largest fireworks display. Along with these massive public fireworks displays, smaller, illegal pyrotechnics are sometimes let off on the lawns of private residences. Considering the scant research on the effects of pyrotechnics on human health, as fire-bursting occurrences occur more frequently and with greater intensity, this societal issue must be addressed.

Air pollution, noise, and poor visibility are the main consequences observed after coordinated fireworks (Lin 2016). Yao et al. (2019) have looked at the health problems brought on by the poor air quality environment during the Chinese Spring celebration in the hotspots of Shanghai from 2013 to 2017. The study found a significant decrease in mortality after the stringent regulations enforced on fireworks displays. The burning of bonfires and the setting of fireworks on 5 November each year mark Guy Fawkes Night, which is a joyous festival in the United Kingdom. Hamad et al. (2016) investigated increased concentrations of metal dust and $PM_{2.5}$-PTE (potentially toxic elements) which were observed in the United Kingdom and linked to lung cancer in both children and adults during this festival at many hotspots in central London. A first report on the microplastic abundance in the Thames following a 2020 New Year's fireworks display (Devereux et al. 2022) in London shows the identification of 2760 pieces of microplastic (99% fibres) in Westminster on the Thames. Likewise, $PM_{2.5}$ peaks were observed and reported by Seidel and Birnbaum (2015) in the United States, between 9:00 and 10:00 pm on 4 July, in 315 stations, and noticed the decline at noon on 5 July, reflecting the New Year celebrations in the United States. This environment with poor air quality produced health issues including severe lung disorders. For the seventh year in a row since 2014, Budapest, Hungary, estimated the largest rate of surface density deposition of smoke in segmental and sub-segmental bronchi, representing an excessive risk to human health during the fireworks related to St. Stephen's Day (Salma et al. 2023). These results point to the adverse impacts of air pollution where children were more susceptible with a three times higher surface density deposition rate in the trachea of children compared to adults.

The rich and varied cultural legacy of India makes it a land of festivals and celebrations. Indian festivals have vibrant celebrations and impressive fireworks that contribute to air pollution. The commonly known national festivals of India are Holi and Diwali. Holi is a summer festival observed by releasing coloured powders into the air, which increases the number of microscopic particles in the atmosphere. The festival of lights known as Diwali, typically takes place between October and November. Owing to the large explosions of firecrackers during this festival, harmful gases and toxic substances are released into the atmosphere, loud noises from speakers and firecrackers, and dry waste, causing health problems for children, patients, and the elderly. It was observed that air pollution was much higher during Diwali in Delhi than on other days and a causal assessment of the contribution of burning Diwali firecrackers to air pollution was investigated (Ghei and Sane 2018). Several studies have been conducted in different parts of India to assess the impact of fireworks on PM10-related air quality (Ravindra et al. 2022). Below is a detailed analysis of toxic air pollutants observed after the fireworks display during the Vishu festival in Kerala, providing a prime example of firework-related toxicity and health risk.

7.4.2 A Case Study of the Detrimental Impacts of Air Pollution from Fireworks on Health at Kannur, India

The increased number of PM and the concentration of surface ozone (O_3) were observed at night by pyrotechnics in the Kannur district of Kerala (Nishanth et al. 2012). The Kannur site is a rural area heavily influenced by the marine environment and lacking in industrial activity. Therefore, the background air pollution is relatively low, and the site is highly vulnerable to air pollution, making it easier to detect changes in pollution levels (Resmi et al. 2021). The major highlight of this study is the first documented observation of elevated surface O_3 resulting from its precursor gases released during night-time fireworks. Surface ozone (O_3) is a secondary air pollutant with a high capacity for oxidation and significant potential for global warming (Resmi et al. 2021; Nishanth et al. 2014). At the same time, inhalation of O_3, a powerful oxidizing agent, can lead to oxidative stress in the lungs (Kelly 2003), and direct exposure to atmospheric pollution causes airway epithelial barrier dysfunction through various processes, such as oxidative stress, amplified cytokine reactions, and compromised host defence, all of which contribute to the onset of asthma and COPD (Mahyar et al. 2022) (Fig. 7.1).

Figure 7.2 highlights the difference in the production rate of O_3 levels observed during the night-time and dawn phases of the fireworks from 2015 to 2022, and the decrease in O_3 in 2020 is a result of the restrictions imposed on the fireworks following the lockdown measures implemented to combat the COVID-19 pandemic. Owing to the larger number of fires during the night-time, O_3 production is more significant during this phase compared to the early morning phase, as shown in Fig. 7.2. The rise in O_3 concentration observed is attributed to the favourable atmospheric conditions that promote O_3 production at this specific location. It is also noted that the slow production of O_3 and particulates in the late-night hours can trap air pollutants, posing a high risk of respiratory disease. This presents an opportunity to examine the intersection of atmospheric chemistry associated with the sudden increase in nocturnal air pollution. The variation in air quality linked to fireworks is depicted in Fig. 7.3 (Resmi et al. 2019).

An attempt to investigate the number of cardiorespiratory cases caused by air pollution from firecrackers has been launched with the support of the Government Hospital Asthma Clinic, Kannur, and the preliminary results of the investigation

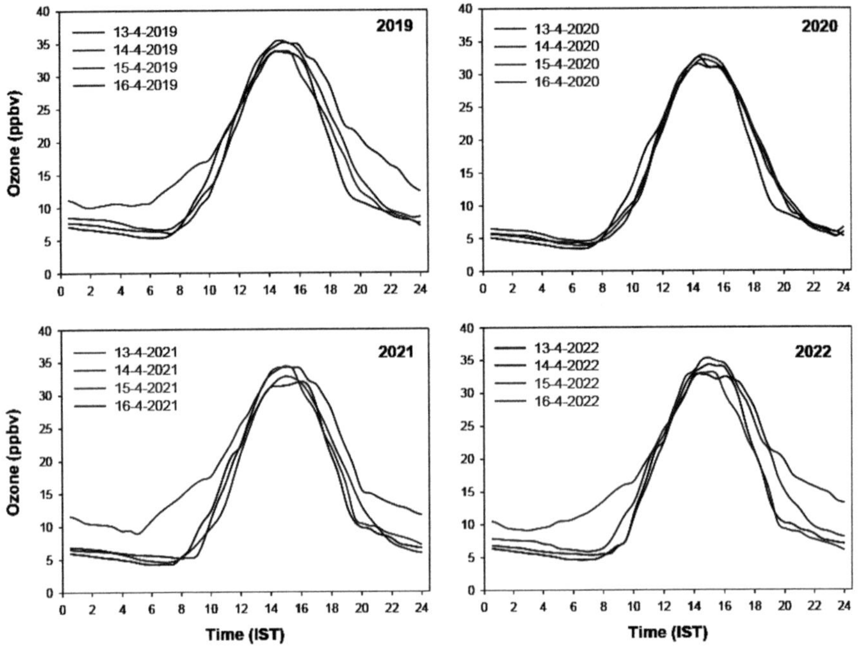

Fig. 7.1 Enhancement of surface O_3 in the study period. (By Resmi et al. 2019)

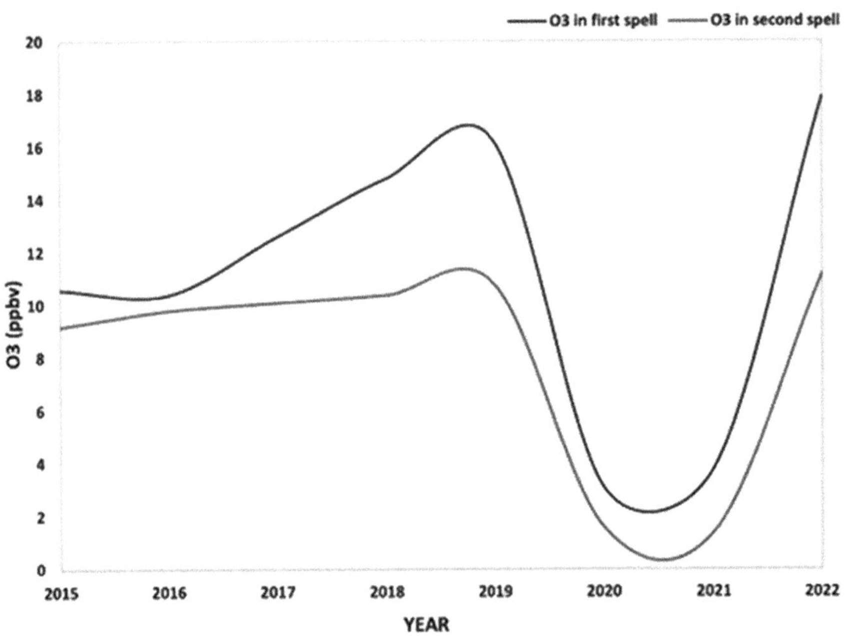

Fig. 7.2 Increasing trend of O_3 production in the two different spells of fireworks

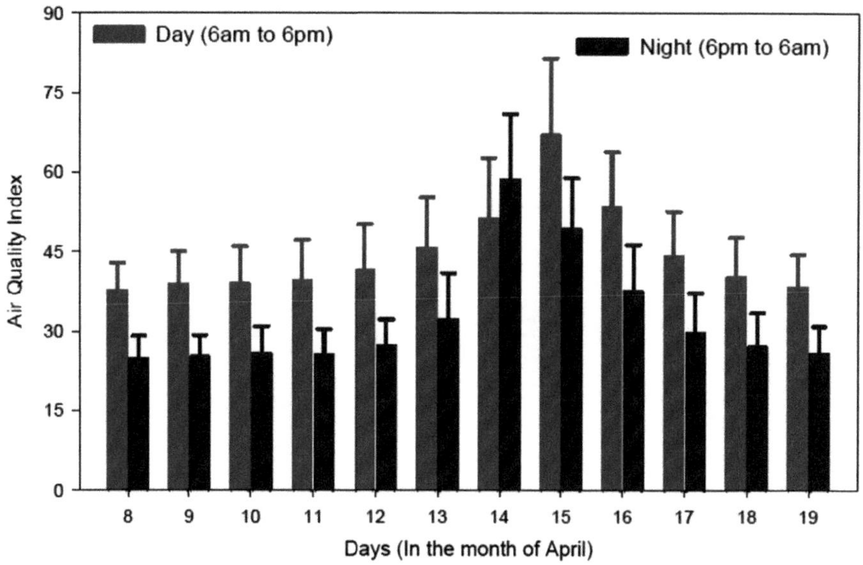

Fig. 7.3 Variation of Air Quality Index (AQI) in Vishu days. (By Resmi et al. 2019)

have been published (Kumar et al. 2022). The Asthma Clinic is primarily a network of such clinics in all private and government hospitals within an 8 km radius to collect details of existing and new cases of respiratory patients after excluding smokers. Figure 7.4 shows the significant increase in the number of cases before and after the fireworks in April, in which Vishu is celebrated in the middle of this month.

The data are categorized into four age groups (3–15, 16–30, 31–55, and 55+) to analyse the striking differences in health issues in each group. Difficulty breathing and wheezing are significantly more pronounced in the 16–30 age group (almost nine- and fivefold increases, respectively), as they are the group most enthusiastic about firefighting. They are extremely vulnerable to the smoke, which causes severe breathing difficulties after the party. Due to sensitivity and the weakened immune system, wheezing is also more common in the over-55 age group. This is also illustrated by the growing number of cases of coughing in the 31 to 55 age group, which is mainly due to the poor quality of indoor air. The increase in new cases reported within a week shows the severe inflammation caused by the powerful firecrackers, increasing susceptibility to respiratory infections.

Numerous nations have implemented restrictions on the use of firecrackers during festivals. China has gone so far as to prohibit the use of firecrackers in many of its cities. Similarly, Germany, France, and the city of Brampton in Canada have banned the sale of fireworks during the festival season. In India, the city of Pune has issued an ordinance outlining regulations for the sale and use of fireworks within its jurisdiction during Diwali. Considering the poor air quality in Delhi during Diwali, the Delhi government has enacted a complete ban on the production, sale, and use

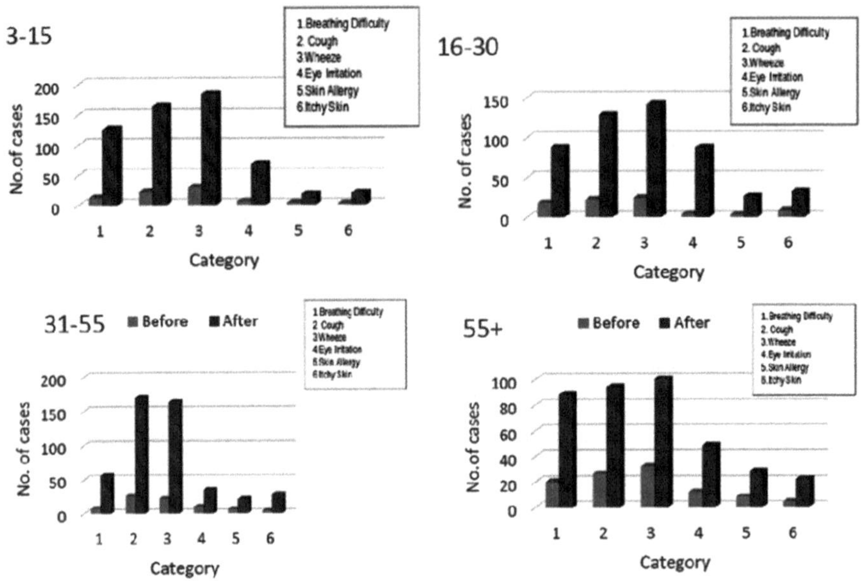

Fig. 7.4 The average number of observed cases for different age groups and disease categories. The average number of observed cases for different age groups and disease categories. (Adapted from Kumar et al. 2022)

of all types of fireworks until the conclusion of the festival. This practice has been in place for 2 years, in 2021 and 2022, due to concerns surrounding environmental pollution as a social issue. The Delhi High Court has also prohibited the sale of fireworks under Article 25 of the Indian Constitution, considering scientific research and the potential hazards associated with pyrotechnics.

7.5 Conclusion

Global warming is caused by the uncontrolled emission of air pollutants into the atmosphere, and this results in a thermal energy imbalance in the atmosphere and oceans, which influences a change in the air and ocean circulation pattern and ultimately impacts the climate. Owing to unexpected floods, droughts, poor air quality, and other environmental catastrophes, climate change poses serious risks to human health. Being one of the most densely populated nations in the world, India has been especially vulnerable to the harmful health consequences of air pollution and climate change. Rapid urbanization, quick expansion, and lack of concern for the environmental effects of festivals lead to air pollution, which eventually has detrimental effects on both human health and climate change.

Festival fireworks are a popular, attractive event and are often perceived as a visually dazzling play of colours in the sky. However, they contribute to air pollution, which has harmful effects on human health and the environment. Studies on short-term anthropogenic exposure to nocturnal pyrotechnics and the effects of firework emissions on air quality and health are relatively scarce. The emission of toxic air pollutants during the fireworks and the resulting poor air quality are analysed during the fireworks at the Vishu festival in Kannur, Kerala, to illustrate the health issues as a case study. For this reason, the massive emissions from fireworks are underestimated, especially given India's large population and lively festival celebrations. Low- and middle-income countries are particularly vulnerable to health problems caused by air pollution and climate change. So, they need to be equipped and trained to protect themselves from health hazards.

Acknowledgements SS wishes to acknowledge the unwavering support received from Mohomed Quiumi, Senior Engineering Manager, and Richard Owusu, Engineering Manager, Control and Protection Systems, GE Vernova, UK. SK would like to thank Professor Santhosh Chidangil and all faculty members of DAMP, MAHE, for their support and would like to thank Dr. Rahul of PSPH, MAHE, for his constant inspiration in making this work a reality.

References

Abbass K, Qasim MZ, Song H, Murshed M, Mahmood H, Younis I. A review of the global climate change impacts, adaptation, and sustainable mitigation measures. Environ Sci Pollut Res. 2022;29(28):42539–59. https://doi.org/10.1007/S11356-022-19718-6.

Bellouin N, Davies W, Shine P, Quaas J, Mülmenstädt J, Forster P, Smith C, Lee L, Regayre L, Brasseur GP, Sudarchikova N, Bouarar I, Boucher O, Myhre G. Radiative forcing of climate change from the Copernicus reanalysis of atmospheric composition. Earth Syst Sci Data. 2020;12(3):1649–77. https://doi.org/10.5194/ESSD-12-1649-2020.

Blenkinsop S, Alves LM, Smith AJP. ScienceBrief review: climate change increases extreme rainfall and the chance of floods. In: Critical issues in climate change science; 2021. p. 1–5. https://sciencebrief.org/uploads/reviews/ScienceBrief_Review_RAINFALL_Jun2021.pdf.

Brasseur GP, Schultz M, Granier C, Saunois M, Diehl T, Botzet M, Roeckner E, Walters S. Impact of climate change on the future chemical composition of the global troposphere. J Clim. 2006;19(16):3932–51. https://doi.org/10.1175/JCLI3832.1.

Brunekreef B, Holgate ST. Air pollution and health. Lancet. 2002;360(9341):1233–42. https://doi.org/10.1016/S0140-6736(02)11274-8.

Cramer W, Guiot J, Fader M, Garrabou J, Gattuso JP, Iglesias A, Lange MA, Lionello P, Llasat MC, Paz S, Peñuelas J, Snoussi M, Toreti A, Tsimplis MN, Xoplaki E. Climate change and interconnected risks to sustainable development in the Mediterranean. Nat Clim Chang. 2018;8(11):972–80. https://doi.org/10.1038/s41558-018-0299-2.

Crutzen PJ, Heidt LE, Krasnec JP, Pollock WH, Seiler W. Biomass burning as a source of atmospheric gases CO, H2, N2O, NO, CH3Cl and COS. Nature. 1979;282(5736):253–6. https://doi.org/10.1038/282253a0.

David LM, Ravishankara AR, Kodros JK, Pierce JR, Venkataraman C, Sadavarte P. Premature mortality due to PM2.5 over India: effect of atmospheric transport and anthropogenic emissions. GeoHealth. 2019;3(1):2–10. https://doi.org/10.1029/2018GH000169.

Devereux R, Westhead EK, Jayaratne R, Newport D. Microplastic abundance in the Thames River during the New Year period. Mar Pollut Bull. 2022;177:113534. https://doi.org/10.1016/J.MARPOLBUL.2022.113534.

Durand G, van den Broeke MR, Le Cozannet G, Edwards TL, Holland PR, Jourdain NC, Marzeion B, Mottram R, Nicholls RJ, Pattyn F, Paul F, Slangen ABA, Winkelmann R, Burgard C, van

Calcar CJ, Barré JB, Bataille A, Chapuis A. Sea-level rise: from global perspectives to local services. Front Mar Sci. 2022;8:709595. https://doi.org/10.3389/FMARS.2021.709595/BIBTEX.

Fuller R, Landrigan PJ, Balakrishnan K, Bathan G, Bose-O'Reilly S, Brauer M, Caravanos J, Chiles T, Cohen A, Corra L, Cropper M, Ferraro G, Hanna J, Hanrahan D, Hu H, Hunter D, Janata G, Kupka R, Lanphear B, et al. Pollution and health: a progress update. Lancet Planetary Health. 2022;6(6):e535–47. https://doi.org/10.1016/S2542-5196(22)00090-0.

Ghei D, Sane R. Estimates of air pollution in Delhi from the burning of firecrackers during the festival of Diwali. PLoS One. 2018;13(8):e0200371. https://doi.org/10.1371/JOURNAL.PONE.0200371.

Goswami BN, Chakraborty D, Rajesh PV, Mitra A. Predictability of South-Asian monsoon rainfall beyond the legacy of Tropical Ocean Global Atmosphere program (TOGA). Npj Clim Atmos Sci. 2022;5(1):1–12. https://doi.org/10.1038/s41612-022-00281-3.

Gurjar BR, Ravindra K, Nagpure AS. Air pollution trends over Indian megacities and their local-to-global implications. Atmos Environ. 2016;142:475–95. https://doi.org/10.1016/J.ATMOSENV.2016.06.030.

Hamad S, Green D, Heo J. Evaluation of health risk associated with fireworks activity at Central London. Air Qual Atmos Health. 2016;9(7):735–41. https://doi.org/10.1007/S11869-015-0384-X/FIGURES/3.

IPCC. Global warming of 1.5 °C. In: Masson-Delmotte V, Zhai P, Pörtner H-O, Roberts D, Skea J, Shukla PR, Pirani A, Moufouma-Okia W, Péan C, Pidcock R, Connors S, Matthews JBR, Chen Y, Zhou X, Gomis MI, Lonnoy E, Maycock T, Tignor M, Waterfield T, editors. An IPCC special report on the impacts of global warming of 1.5°C. Cambridge University Press; 2018. p. 3–24. https://doi.org/10.1017/9781009157940.001.

Jabbar HK, Hamoodi MN, Al-Hameedawi AN. Urban heat islands: a review of contributing factors, effects and data. IOP Conf Ser Earth Environ Sci. 2023;1129(1):012038. https://doi.org/10.1088/1755-1315/1129/1/012038.

Kelly FJ. Oxidative stress: its role in air pollution and adverse health effects. Occup Environ Med. 2003;60(8):612–6. https://doi.org/10.1136/OEM.60.8.612.

Kumar SM, Nishanth T, Valsaraj KT. Association of Increased Cardiorespiratory Cases with the mounting levels of air pollution caused by fireworks. Open J Cardiol Heart Dis. 2022;3(5):1–4. https://doi.org/10.31031/OJCHD.2022.03.000574.

Lal S, Naja M, Subbaraya BH. Seasonal variations in surface ozone and its precursors over an urban site in India. Atmos Environ. 2000;34(17):2713–24. https://doi.org/10.1016/S1352-2310(99)00510-5.

Lin CC. A review of the impact of fireworks on particulate matter in ambient air. J Air Waste Manag Assoc. 2016;66(12):1171–82. https://doi.org/10.1080/10962247.2016.1219280/SUPPL_FILE/UAWM_A_1219280_SM2186.PDF.

Mahyar A, Ubags ND, Bruder D, Hiemstra PS, Sidhaye V, Rezaee F, Heijink IH. Role of air pollutants in airway epithelial barrier dysfunction in asthma and COPD. Eur Respir Rev. 2022;2022(31):210112. https://doi.org/10.1183/16000617.0112-2021.

Mc Evoy D, Buggy CJ. Respiratory effects of running in urban areas with high and low ambient air pollution: a scoping review. Sustain Environ. 2023;9(1) https://doi.org/10.1080/27658511.2023.2194493/SUPPL_FILE/OAES_A_2194493_SM9448.DOCX.

Molina MJ, Molina LT. Megacities and atmospheric pollution. J Air Waste Manage Assoc. 2004;54(6):644–80. https://doi.org/10.1080/10473289.2004.10470936.

Nishanth T, Praseed KM, Rathnakaran K, Satheesh Kumar MK, Ravi Krishna R, Valsaraj KT. Atmospheric pollution in a semi-urban, coastal region in India following festival seasons. Atmos Environ. 2012;47:295–306. https://doi.org/10.1016/J.ATMOSENV.2011.10.062.

Nishanth T, Praseed KM, Kumar MKS, Valsaraj KT. Influence of ozone precursors and PM10 on the variation of surface O3 over Kannur, India. Atmos Res. 2014;138:112–24. https://doi.org/10.1016/J.ATMOSRES.2013.10.022.

Noël C, Vanroelen C, Gadeyne S. Qualitative research about public health risk perceptions on ambient air pollution. A review study. SSM Popul Health. 2021;15:100879. https://doi.org/10.1016/J.SSMPH.2021.100879.

Ramanathan V. Climate change, air pollution, and health: common sources, similar impacts, and common solutions. In: Health of people, health of planet and our responsibility: climate change, air pollution and health; 2020. p. 49–59. https://doi.org/10.1007/978-3-030-31125-4_5/FIGURES/1.

Ravindra K, Kumar S, Mor S. Long term assessment of firework emissions and air quality during Diwali festival and impact of 2020 fireworks ban on air quality over the states of Indo Gangetic Plains airshed in India. Atmos Environ. 2022;285:119223. https://doi.org/10.1016/J.ATMOSENV.2022.119223.

Resmi CT, Nishanth T, Satheesh Kumar MK, Balachandramohan M, Valsaraj KT. Temporal changes in air quality during a festival season in Kannur, India. Atmosphere. 2019;10(3):137. https://doi.org/10.3390/ATMOS10030137.

Resmi CT, Ye F, Satheesh S, Nishanth T, Satheesh Kumar MK, Balachandramohan M, Manivannan D, Hu J, Valsaraj KT. Variation of trace gases in Kannur Town, a coastal South Indian city. Environ Challeng. 2021;5:100336. https://doi.org/10.1016/J.ENVC.2021.100336.

Saha L, Kumar A, Kumar S, Korstad J, Srivastava S, Bauddh K. The impact of the COVID-19 lockdown on global air quality: a review. Environ Sustain. 2022;5(1):5–23. https://doi.org/10.1007/S42398-021-00213-6.

Saikawa E, Panday A, Kang S, Gautam R, Zusman E, Cong Z, Somanathan E, Adhikary B. Air pollution in the Hindu Kush Himalaya. In: The Hindu Kush Himalaya assessment: mountains, climate change, sustainability and people; 2019. p. 339–87. https://doi.org/10.1007/978-3-319-92288-1_10/FIGURES/18.

Salma I, Farkas Á, Weidinger T, Balogh M. Firework smoke: impacts on urban air quality and deposition in the human respiratory system. Environ Pollut. 2023;328:121612. https://doi.org/10.1016/J.ENVPOL.2023.121612.

Santer BD, Po-Chedley S, Zhao L, Zou CZ, Fu Q, Solomon S, Thompson DWJ, Mears C, Taylor KE. Exceptional stratospheric contribution to human fingerprints on atmospheric temperature. Proc Natl Acad Sci USA. 2023;120(20):e2300758120. https://doi.org/10.1073/PNAS.2300758120/SUPPL_FILE/PNAS.2300758120.SAPP.PDF.

Seidel DJ, Birnbaum AN. Effects of Independence Day fireworks on atmospheric concentrations of fine particulate matter in the United States. Atmos Environ. 2015;115:192–8. https://doi.org/10.1016/J.ATMOSENV.2015.05.065.

Shubham S, Zhang M, Anshika JG, Zhang H, Kota SH. Effect of restricted emissions during COVID-19 on air quality in India. Sci Total Environ. 2020;728:138878. https://doi.org/10.1016/J.SCITOTENV.2020.138878.

Shukla S. Winter air pollution and stubble burning in North India: a regulatory governance perspective | ORF. Observer Research Foundation, 2021, November 20. https://www.orfonline.org/expert-speak/winter-air-pollution-and-stubble-burning-in-north-india/.

Solomon S, Plattner GK, Knutti R, Friedlingstein P. Irreversible climate change due to carbon dioxide emissions. Proc Natl Acad Sci USA. 2009;106(6):1704–9. https://doi.org/10.1073/PNAS.0812721106/SUPPL_FILE/0812721106SI.PDF.

Swapna P, Ravichandran M, Nidheesh G, Jyoti J, Sandeep N, Deepa JS, Unnikrishnan AS. Assessment of climate change over the Indian region: a report of the ministry of earth sciences (MOES), government of India. In: Krishnan R, Sanjay J, Gnanaseelan C, Mujumdar M, Kulkarni A, Chakraborty S, editors. Assessment of climate change over the Indian region: a report of the Ministry of Earth Sciences (MoES), Government of India. Singapore: Springer; 2020. p. 1175–89. https://doi.org/10.1007/978-981-15-4327-2_9.

Tetali S, Baird N, Klima K. A multicity analysis of daytime Surface Urban Heat Islands in India and the US. Sustain Cities Soc. 2022;77:103568. https://doi.org/10.1016/J.SCS.2021.103568.

Topál D, Ding Q. Atmospheric circulation-constrained model sensitivity recalibrates Arctic climate projections. Nat Clim Chang. 2023;13(7):710–8. https://doi.org/10.1038/s41558-023-01698-1.

Wang H, Lu X, Jacob DJ, Cooper OR, Chang KL, Li K, Gao M, Liu Y, Sheng B, Wu K, Wu T, Zhang J, Sauvage B, Nédélec P, Blot R, Fan S. Global tropospheric ozone trends, attributions, and radiative impacts in 1995-2017: an integrated analysis using aircraft (IAGOS) observations, ozonesonde, and multi-decadal chemical model simulations. Atmos Chem Phys. 2022;22(20):13753–82. https://doi.org/10.5194/ACP-22-13753-2022.

Yao L, Wang D, Fu Q, Qiao L, Wang H, Li L, Sun W, Li Q, Wang L, Yang X, Zhao Z, Kan H, Xian A, Wang G, Xiao H, Chen J. The effects of firework regulation on air quality and public health during the Chinese Spring Festival from 2013 to 2017 in a Chinese megacity. Environ Int. 2019;126:96–106. https://doi.org/10.1016/J.ENVINT.2019.01.037.

Zhang B, Tian H, Ren W, Tao B, Lu C, Yang J, Banger K, Pan S. Methane emissions from global rice fields: magnitude, spatiotemporal patterns, and environmental controls. Glob Biogeochem Cycles. 2016;30(9):1246–63. https://doi.org/10.1002/2016GB005381.

Heat Health and Climate Change: Governance in India

8

Alyssa Selma Martins, Neethi V. Rao, and Abhiyant Tiwari

Abstract

Heat is one of the fastest-growing weather hazards, affecting health, ecosystems, and livelihoods. With climate change intensifying, heat waves are bound to increase causing disastrous consequences. This chapter explores the relationship of heat on health in India, with a special focus on its impact on the health of people, living conditions, and means of living. A global contextualization is provided to aid with the impact of rising temperatures on the area of focus, that is, India. It examines the repercussions of heat and analyses the role of existing policies such as Heat Action Plans (HAPs), in addressing the same.

Keywords

Heat waves · Climate change · Livelihood · India · Heat stress · Heat Action Plans (HAPs) · Energy

A. S. Martins (✉)
Prasanna Schools of Public Health, Manipal Academy of Higher Education (MAHE), Manipal, India
e-mail: alyssa.martins@manipal.edu

N. V. Rao
CSEP Research Foundation, New Delhi, India

A. Tiwari
Climate Resilience & Health Consultant, NRDC, Guwahati, India

H. Brand et al. (eds.), *Climate Change and Public Health*,
https://doi.org/10.1007/978-981-95-0895-2_8

8.1 Introduction

With every passing day, temperature records are being shattered across the globe. Extreme heat events are becoming a common reoccurrence driven by climate change (Carbonari et al. 2019; Jairaj and Malaviya 2019) (Miller et al. 2021; Somerville et al. 2012). The surge in heat waves is attributed to the rise in consumption of fossil fuels such as coal, oil, and gas that lead to emissions of GHGs trapping the heat in the earth's atmosphere. El Niño, a weather phenomenon where there is a rise in ocean temperatures specifically in the Pacific Ocean, further interacts with global warming to increase the frequency and intensity of heat waves. Together, they were considered the chief reasons behind the first week of July 2023 as the hottest week on record. Excessive fossil fuel consumption combined with climatic events like El Niño has caused events such as ocean warming, record low levels of Antarctic Sea ice, droughts, flooding, and heat waves (Plumer and Shao 2023). Extreme heat affects various aspects of the ecosystem and increases the risk of disasters that affect health tremendously. Rising temperatures can worsen droughts and contribute to conditions causing wildfires. The ecological disruptions include damage to crop yield impacting food security and the spread of VBDs. These situations also cause displacement of people leading to the condition of climate refugees (Centers for Disease Control and Prevention 2008).

Human health is vulnerable to changes in weather patterns and other factors of climate change. The impact is in two dimensions, that is, exacerbating existing health issues affecting mortality and morbidity and the emergence of new conditions that will potentially affect areas that were unaffected previously such as the spread of waterborne and VBDs (Rocklöv and Dubrow 2020; Walker 2018). High temperatures affect human health, affect mortality, harm pregnancy, and negatively affect mental health (Ebi et al. 2021). The body's capacity to control its internal temperature is compromised by temperature extremes, which have a direct impact on health. Heat cramps, heat exhaustion, heatstroke, and hyperthermia from extremely hot weather events can all be caused by a loss of internal temperature regulation. Heat-associated temperature extremes can aggravate several chronic illnesses, including those connected to diabetes, cardiovascular disease, respiratory disease, and cerebrovascular disease (NIH 2018). Considering the dire consequences, the NDMA has released guidelines on first aid for heat disorders to counter the effects of heat (National Disaster Management Authority et al. 2019). High ambient temperatures are also influenced by external factors such as 'low vegetation cover', land degradation, and high population density along with indoor heat generated by building style and materials, ventilation, and cooling methods (Tasgaonkar et al. 2022).

In India, living conditions are expected to be as hot as the Sahara Desert, given current emission rates (IPCC). Accelerated urbanization has compounded this risk through urban heat island effects and high population density. Urban heat waves are increasingly more frequent, more intense, and of longer durations. Indian cities are on average two degrees hotter than rural areas (Balasubramanian 2020). About 483 million people live in these cities, of whom about 63 million live in low-income neighbourhoods.

Acknowledging the severity of the situation, the GoI has taken numerous measures to address and mitigate the effects of this prolonged natural disaster. Post-2015, the central government changed its approach from 'response-centric' to 'risk reduction', that is, working towards the holistic management of disasters. It led to a successful reduction in deaths related to heat 2040 deaths in 2015 to 4, 5 years later (National Disaster Management Authority and Ministry of Home Affairs n.d.).

Extreme heat affects the health, comfort, livelihood, and well-being of the population. This chapter discusses the impact of heat on vulnerable people with a focus on India. It outlines the impact of extremely high temperatures on health, living conditions, and economic conditions including livelihood. Through the analysis of existing policies, this chapter examines the effectiveness of the policies and provides recommendations to address this dire issue.

8.2 Extreme Heat and the Vulnerable

Due to climate change, the number of people affected by heat waves is increasing. According to the WHO, people who experienced heat waves increased by over 125 million between 2000 and 2016 (WHO 2018). The Billion Dollar Weather Catastrophes database compiled by the National Oceanic and Atmospheric Administration reveals that over 600 people each year on average died from 1999 to 2009 from extreme heat-related causes, and heat is one of the main causes of weather-related deaths in the United States (C2ES. n.d.; Dahl et al. 2019). Likewise, the Russian Federation witnessed a 44-day heat wave in 2010 that resulted in 56,000 deaths (WHO 2018). Recently, in Kharghar, Navi Mumbai, Maharashtra, 14 people perished from heatstroke while attending a formal government function that was conducted in an open field on 16 April 2023 (Koll 2023).

Heat stress and hot environmental conditions can worsen mental health, escalate heat-related deaths and existing diseases, and increase unfavourable pregnancy outcomes as outlined below. Recent studies show that most people can adapt biologically and physically to small changes in average normal temperatures. However, children, pregnant women, people with chronic health conditions, and older persons are more vulnerable to negative consequences because they are less able to regulate their body temperatures and adapt to changes in ambient heat (Ebi et al. 2021; NIH 2018). Other at-risk categories include individuals who work outdoors, outdoor athletes, people who are socially isolated or with low incomes, as well as people who live in communities of colour, highlighting that social vulnerabilities can exacerbate the climatological risk (NIH 2018). In infants, heat-related illnesses range from mild rashes to life-threatening cases of organ failure due to heatstroke (Benz and Burney 2021). Heat waves also increase the risk of preterm birth (PTB) which harms the health of the infants. Pregnant women may experience difficulties during heat waves because they are more susceptible to heat-related illnesses, which can cause issues during pregnancy such as preterm labour, preeclampsia, and neural tube abnormalities. Hot weather-related dehydration can result in preterm labour, premature birth, and low birth weight in babies (Chersich et al. 2020; Samuels et al.

2022). Extreme heat exposure during pregnancy can raise the risk of preeclampsia, gestational diabetes, and other pregnancy-related disorders that can have an impact on both the mother's and the unborn child's health in the long run (Chersich et al. 2020, 2023; Dalugoda et al. 2022). Increases in heat-related ailments, such as cardiovascular and respiratory problems, kidney disease, and respiratory issues, are also other illnesses that are linked to extremely hot days (Ebi et al. 2021; Knowlton n.d.). For those without access to cooling, the heat during the night can be uncomfortable and create health issues. High ambient temperatures escalate existing mental health conditions, for example, post-traumatic stress disorder and bipolar disorder, cause mood swings, and lead to an increase in depression episodes and suicide (Liu et al. 2021).

8.3 Livelihood and Mitigation

A report by the International Labour Organization (ILO) draws attention to the projection that heat stress could result in the loss of 34 million full-time jobs in India by 2030 (Kjellstrom et al. 2019). The construction industry will also see an increase in lost working hours because of heat stress, which will affect both male and female workers but will have the greatest impact on the agriculture sector. In addition to lowering physical labour capacity and motor-cognitive performance, which has an impact on productivity, high heat stress also raises the possibility of occupational health issues (PTI 2019). According to a 2021 study, outdoor workers in developing nations have higher core body temperatures than indoor employees and are two to three times more likely to become dehydrated, which increases the risk of kidney damage and other illnesses associated with dehydration. Heat-related illnesses (HRIs), which can happen when the human body absorbs and emits more heat than it can dissipate, are a potential risk for workers in many indoor and outdoor settings. When heat intake exceeds heat loss during intense physical exertion in a warm environment, the core temperature increases quickly (Venugopal et al. 2021). As noted in the IPCC report, poor and marginalized populations have been directly impacted by climate change due to its effects on their livelihoods, such as crop yield losses, home destruction, and food insecurity, as well as indirectly due to higher food costs (Pachauri et al. 2014). When global warming reaches a critical point, it will harm the health and financial security of farm labourers, daily wage employees, and other vulnerable communities. Such groups are also more susceptible to dehydration, heat cramps, and heat exhaustion, which, if left untreated, can progress to more serious conditions, especially during heat waves (Kjellstrom et al. 2016, 2018).

Sudden or progressive change in place of habitat due to climate change can give rise to climate migrants that may shift temporarily or permanently within their country or across international borders. Factors such as extreme weather events like rising temperatures or flooding, loss of livelihood (damage to agricultural land), and sea level rise may contribute to the movement of people. According to the IMDC report, 32.6 million (326 lakhs) internal displacements occurred in 2022 as a result of disasters (IDMC 2023). And, it is projected to rise by 2050 (Clement et al. 2021).

The number can be higher than estimated as specific data on heat-related migration is not available as unlike disasters, this kind of migration is not time-restricted. Excessive heat is likely to have been a major factor in combination with other socio-economic factors like livelihoods influencing the movement of people.

India is already facing the pressure of climate migration. The Sundarbans in Bangladesh is a risk area due to its vanishing islands. This has created an outflux of migrants to safe areas. It is estimated that around 50–120 million people will migrate to the neighbouring country, India (Sawant and Sanjeev 2022). Internally, environmental disasters led to the displacement of around 3,856,000 people in 2020. Events such as sea level rise, cyclones, and flooding are becoming common occurrences. The latter is projected to occur in areas that did not flood earlier. It also has a historical perspective on the flooding of the Ganga and Brahmaputra rivers (Abdulali and Abdulali 2021). On the other hand, Satbhaya in Odisha faced coastal erosion with only four villages remaining on the surface while three submerged. The people were relocated to Bagapatia in Kendrapara district which will be the first resettlement colony in India for people affected by climate change (Rathore 2023). Climate migration along with other factors can exacerbate the effects of heating such as by contributing to urban heat island effects. Climate extremes, at the destination region, can also negatively affect these migrants even though migration is seen as a key adaptation strategy to mitigate the adverse effects of climate change, for example, rising sea levels and flooding in Mumbai. According to the 2011 Census, every six urban Indians live in a slum with cramped and unsanitary conditions (Hari et al. 2021). Employers of construction workers, head loaders, and restaurant/hotel staff frequently agree to pay reductions in exchange for their employees' willingness to stay on the job site. Others search for reasonably priced rental homes in slums, mess houses, or worker colonies that are frequently found close to major workplaces. Due to the great demand for this housing, the migrants are forced to pay higher rent while living in congested, filthy conditions without essential water and sanitation amenities. People without housing options sleep on city sidewalks, train platforms, and bus stops and have heightened vulnerability to excessive heat (Akoijam et al. 2017; Madan 2023; Patel 2021).

8.4 Heating and Cooling and Implications for Energy

High indoor temperatures affect 'health, thermal comfort and coping practises' (Tasgaonkar et al. 2022). Similarly, outdoor heat in rural and urban areas plays an important role in influencing the temperatures. Due to 'urban heat islands[1]', the urban climate has higher temperatures that are attributed to numerous variables,

[1] When cities have greater air temperatures than the surrounding countryside, the phenomenon is referred to as places affected by these elevated temperatures also known as the urban heat island effect (EPA 2022).

including reduced radiant heat[2] loss in the urban canopy layer, buildings that can block the flow of wind resulting in slower wind speeds, and greater radiation exposure.[3] The UHI effect, which is brought on by mostly paved surfaces that absorb and radiate heat and the absence of green spaces and tree cover in these places, puts city dwellers at greater risk of heat-related illness (Benz and Burney 2021; Gregory et al. n.d.; Mohajerani et al. 2017) both at high resolution (within cities or counties). Urban India experiences Surface Urban Heat Island Intensity (SUHII) 2 °C higher than its rural counterpart due to the reasons highlighted above.

The building materials capture and trap the heat and release the stored heat at night. This affects the temperature in the location. Meanwhile, the presence of trees, vegetation, and water bodies lowers surface temperatures. This cooling occurs because of transpiration and evaporation. India faces daunting challenges with its unplanned urbanization, poor air quality, and vulnerability to climate threats among other issues. It requires coordination among various actors to mitigate and adapt to climate threats through measures such as green roofs (improve indoor temperature), sustainable buildings (reduce energy consumption which will reduce the heat generated by air conditioning appliances), and trees (boost circulation and improve quality of air), etc. (Nath et al. 2021). A heat map (Fig. 8.1) recording the land and surface temperatures in Delhi captures the cooling effect of trees and water bodies in contrast to the hot areas of the city such as sidewalks and buildings.

Additionally, high temperatures affect the transmission of energy. During heat waves, the ability of transmission lines to carry power may be reduced because of greater summer temperatures, which may result in concerns with electrical reliability such as rolling blackouts. However, in a study based in the USA, it was found that overall energy demand would rise in a warmer future, even though warmer winters will lessen the need for heating. Additionally, as rivers and lakes warm up, their ability to absorb waste heat from power plants decreases. This can lower the thermal efficiency of energy production, making it more challenging for power plants to meet environmental standards regarding the temperature of their cooling water, which may result in plant shutdowns (Dahl et al. 2019), and increase the annual number of days with heat indices exceeding 37.8 °C (100 °F.

Heating and cooling have a tremendous impact on climate. Heating systems consume fossil fuels that release GHGs. Similarly, wood stoves release dangerous indoor and outdoor air pollution together with the GHGs. On the other hand, air conditioning (AC) units are being employed to stay cool indoors. This method consumes a lot of electricity and leaks refrigerants such as hydrofluorocarbons (Norford and Gribkoff n.d.). An analysis by the International Energy Agency draws attention to the current scenario. Nearly, 20 percent of the electricity consumed in buildings

---

[2] Refers to the transfer of heat from a body to a medium. For example, the sun is the source of radiant heat and the urban surface blocks the transfer of heat to the atmosphere increasing temperatures.

[3] Construction materials such as asphalt absorb most of the radiation and emit it as heat thus increasing human exposure to heat.

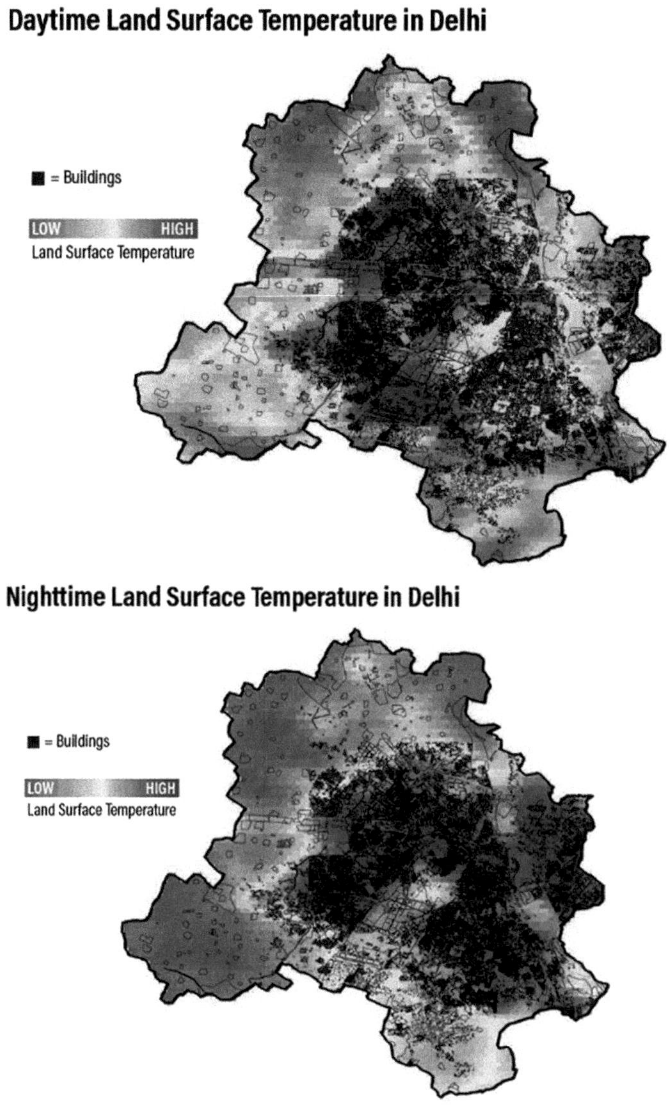

Fig. 8.1 Heat map of Delhi displaying the daytime and night-time temperatures. (By Jairaj and Malaviya 2019; MODIS n.d.)

worldwide today is used by air conditioners and electric fans to stay cool. In addition to increasing emissions, the growing demand for space cooling is placing a tremendous amount of stress on the energy infrastructure of many nations (IEA 2018). The problem is serious in the fastest-growing countries, with the highest growth occurring in countries like India, where the ratio of air conditioning in peak electrical load might increase to 45 percent in 2050 from 10 percent now if no policy action is taken (UNEP and IEA 2020).

IPCC's Sixth Assessment Report (AR 6) stressed the importance of urban green and blue infrastructure. The report stated that measures to mitigate the heat island effect and heat stress include 'green walls, greenways, street trees, urban forests, green roofs, blue spaces, to cool urban areas' (IPCC 2022). By decreasing urban heat, urban trees can help to alleviate some of the effects of climate change. They can also help to minimize storm water runoff, improve air quality, and promote health and well-being.

Efficient cooling has many benefits including protecting the vulnerable from the effects of extreme heat. In India, most of the houses in low-income areas are constructed from heat-trapping materials.[4] In 2009, Mahila Housing Trust (MHT)[5] carried out cooling solutions in Ahmedabad, Gujarat, through a range of initiatives such as green roofs, white paint, Airlite,[6] and ModRoof[7] (K 2023). Also, cool roofs have emerged as one of the most affordable and simple measures to deal with heat. It is demonstrated by the community-led projects in Ahmedabad, Bhopal, Jodhpur, and Surat. In 2021, NRDC[8] along with (MHT) put a solar reflective coating on the rooftops of a few chosen slum residences (P. Madan and Korsh 2021). Cool roofs lower inside temperatures which can also lessen the need for air conditioning. Reflecting sunlight, cool roofs release less heat. Compared to standard roofs, cool roofs can help keep indoor temperatures down by 2 to 5 °C (3.6 to 9 °F), depending on the environment. The Kigali Amendment[9] and the Indian Cooling Action Plan[10] mention cool roofs and energy efficiency measures as critical strategies for sustainable cooling and battling climate change. The advantages of cool roofs also include greater thermal comfort by maintaining lower indoor temperatures, reduced cooling demand in the future from air conditioners, and can help reduce air pollution since they save energy, especially on cooling devices like fans and Acs (Madan and Korsh 2021). 'Cool roofing strategy' was introduced by Telangana state in India in April 2023. The first of its type in India and the world, this policy is implemented to request building permission and safeguard the country's 35 million citizens. Telangana hopes to install 300 square kilometres of cool roofs in the state by 2028. After 5 years of policy implementation, it is anticipated to conserve 30 million tonnes of carbon dioxide, which contributes to global warming, and 600 million units (GWh) of electricity annually (P. Madan et al. 2023).

[4] Houses are built using materials such as tin, asbestos, bricks, cement, concrete, and tarpaulin that trap heat indoors causing great discomfort, affecting health, and lowering productivity.

[5] A nonprofit organization that is working for the betterment of poor communities especially women by improving their urban settlement.

[6] Semi-transparent plastic that allows light to penetrate thus improving the ventilation.

[7] A roof made from environment-friendly materials such as agricultural waste and cardboard.

[8] Natural Resources Defense Council is a non-profit international environmental advocacy group with its headquarters based in the United States.

[9] Amendment (2016) to phase down *hydrofluorocarbons (HFCs) used in refrigeration and air conditioning under the Montreal Protocol (1985).*

[10] Provides a vision for the next 20 years and charts a plan of action to access sustainable cooling.

In 2018, the Draft India Cooling Action Plan (ICAP) was unveiled to address India's expanding cooling needs. The strategy represents an important step in India's efforts to comply with the Montreal Protocol's[11] phase-down of cooling-related chemicals, notably highly polluting hydrofluorocarbons (HFCs) used for air conditioning in vehicles, buildings, and other structures. India is the first nation to create a national strategy to deal with increasing cooling requirements while attempting to lower demand for air conditioning. The ICAP underlines connections between India's cooling program and other efforts and programs dealing with the environment and climate change, including the Mission Innovation,[12] Housing for All,[13] the National Mission on Enhanced Energy Efficiency,[14] and the Smart Cities Mission[15] (Madan 2018).

8.5 India's Policy Response to Excessive Heat

Heat may be efficiently mitigated and managed by breaking down boundaries between urban planning, public health, and emergency management. It requires coordination between various agencies to combat the growing challenges and impacts of heat. A *Nature* article proposes improved heat governance in mitigating and controlling heat as a hazard. It advocates for a dedicated research effort to assist 'heat governance', with a focus on six principles, namely, advance heat equity, mitigate heat, manage risks, develop metrics, coordinate initiatives, and build heat institutions (Keith et al. 2021).

The GoI has taken many steps to deal with the effects of heat waves. In a directive issued in April 2023, The Ministry of Labour and Employment has ordered all states and territories to guarantee readiness and effective management of the impact of the impending heat wave on employees and labourers in various industries. It outlines several tactical actions that must be taken, including rescheduling working hours for employees and workers, ensuring adequate drinking water facilities at workplaces, providing emergency ice packs and heat illness prevention material to construction workers, working with the health department to ensure that workers receive regular health examinations, and adhering to health advisories issued by the Ministry of Health & Family Welfare for Employers and Workers (Madan 2023). Additionally, the letter emphasizes the urgency of giving the Management of Mines orders to take urgent action to provide nearby rest spaces, an appropriate supply of cool water, and electrolyte supplements. Other suggested measures include

[11] It is an international agreement that aims to protect the ozone layer by phasing out the production and consumption of nearly 100 ozone depleting substances (ODS).

[12] A global initiative towards innovation in clean energy to mitigate climate change.

[13] Scheme to provide housing facilities for slum dwellers.

[14] One of the eight national missions under the National Action Plan on Climate Change (NAPCC). It consists of initiatives to enhance energy efficiency in energy-intensive industries.

[15] Program to develop urban areas across the country, making them citizen friendly and sustainable.

allowing work at a slow pace in case the worker feels unwell, allowing rest times and flexible schedules to let workers do the hardest work during the coolest parts of the day, assigning a two-person crew to perform work during extremely hot temperatures, and more (PIB Delhi 2023). Further, to assist state governments and other stakeholders in drafting heat wave management plans, the National Disaster Management Authority (NDMA) has issued National Guidelines on Heat Wave Management.

India has devised Heat Action Plans (HAPs) consisting of four key components—an early warning system to trigger emergency responses, public awareness and community outreach, capacity-building for medical professionals, and finally, the promotion of adaptive measures to reduce heat exposure. It involves governance at every level of government and includes various measures to deal with heat such as preventative actions, emergency responses, and post-heat wave reactions. In May 2010, Ahmedabad experienced a heat wave that caused 1344 deaths prompting the city to develop its HAP. It was the first city in the country to draft a HAP in 2013 along with local and international experts (Climate and Development Knowledge Network (CDKN) 2014). The measures included creating interagency heat action plans; improving capacity at health centres to deal with patients coming with heat-related illnesses; upgrading communication between the authorities and people by taking steps to inform them of upcoming heat events; implementing outreach and training programs for the vulnerable to build resilience, for example, slum communities; and creating awareness and implementing policies to address labour safety (NRDC n.d.). It set an example for drafting HAPs across the country. In India, efforts are coordinated by the IMD and the NDMA to prevent and address heat waves. In 2016, ten cities launched HAPs and the NDMA issued its first heat wave guidelines to provide a framework for the implementation, coordination, and evaluation of extreme heat wave-related activities. NDMA further amended its Heat Guidelines in 2019 to address concerns related to the built environment, including solutions such as the 'cool roofs' highlighted in the previous section (NRDC 2022).

In a bid to develop and deploy HAPs across the country, the Indian government is collaborating with 23 heat-prone states and over 130 towns and districts (NRDC 2022). Although there is a common thread among all HAPs, certain states have unique methods of adapting and mitigating heat waves. For instance, Andhra Pradesh employs media to spread information on awareness regarding the timings when the heat will be at its peak. The state is expected to employ data from the European Centre for Medium-Range Weather Forecasts (ECMWF) (NRDC 2022). In Kerala, green roofs, cool pavements, and incentives for environment-friendly building materials are among the strategies suggested in its HAP. This document was prepared by the Kerala State Disaster Management Authority (KSDMA) to tackle rising temperature levels. Likewise, Maharashtra has taken measures to deal with heat in its state. In 2022, the Brihanmumbai Municipal Corporation (BMC) issued a Climate Action Plan to build Mumbai as a climate-resilient city, to achieve net zero carbon emissions by 2050. Mumbai is the first South Asian metropolis to adopt a net zero goal (World Resources Institute India and C40 Cities Network

2022). BMC has also issued guidelines to the public that include steps to stay hydrated, foods to avoid, drinking buttermilk or oral rehydration salts, and the type of clothing to wear (BMC n.d.; NRDC 2022).

Despite the numerous measures, 14 people died on 16 April 2023 at an event in Navi Mumbai (Koll 2023). This drew attention to the challenges posed by the HAPs. The incident generated a lot of discussion on the HAPs. In his report, Roxy Koll states that the lack of precautions resulted in the deaths. The day of the incident had maximum temperatures between 34 and 38 degrees Celsius with high humidity. The people sat in the open sun with no shed or basic facilities drinking water or washroom facilities (Koll 2023). According to a report by the Centre for Policy Research, HAPs in India suffer from a weak legal framework and are underfunded. The study analysed 37 HAPs and found that it is not drafted with a local context, that is, the plans overlook humidity and warm nights while determining heat thresholds in the region. Second, there is a disconnect between the vulnerable groups identified and the policies drafted to address their issues. Also, each state including the city and district needs to conduct an assessment of the vulnerable population in its region (Pillai and Dalal 2023). In Fig. 8.2, Pillai and Dalal stress the need for the Government and policymakers to focus on existing national and state schemes to ensure proper implementation and integration of HAPs, highlighting the need for intersectoral coordination as a critical aspect of the policy response.

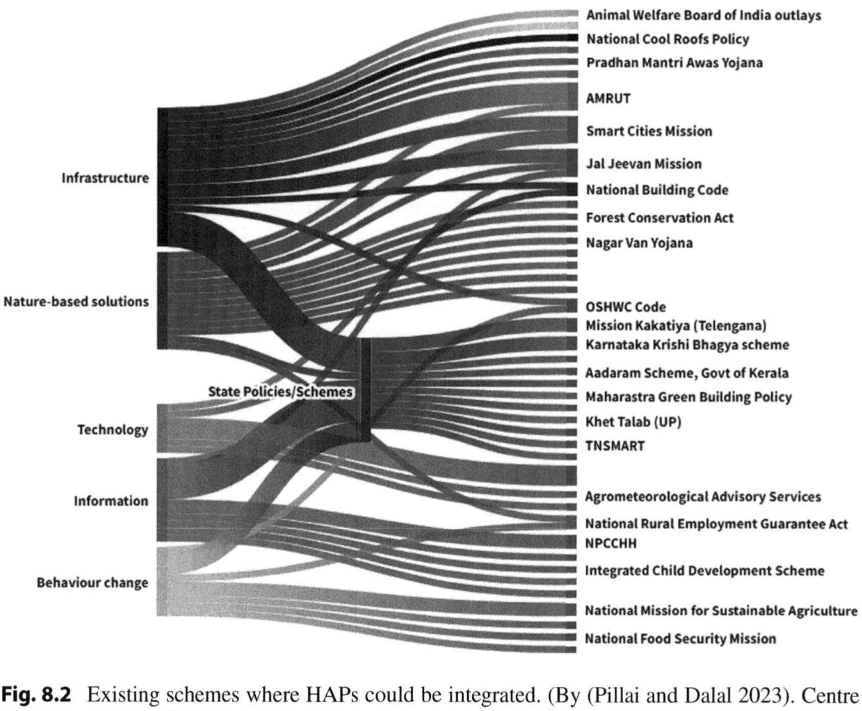

Fig. 8.2 Existing schemes where HAPs could be integrated. (By (Pillai and Dalal 2023). Centre for Policy Research)

8.6 Conclusion

The rising threat of heat is severely affecting the health of the population. Numerous sections of the general population are affected due to the differences in their bodies' ability to regulate temperature. Occurrences of heat waves cause a domino effect affecting various sectors of life such as the destruction of crops that threaten food security, rising prices of agricultural products, and the well-being of daily wage labourers. Various measures are being taken on the international level as well as national level to adapt and mitigate the heat. The UNEP has promised a Global Cooling Pledge and a Cool COP Menu of Actions, ensuring that sustainable cooling will be a top priority at this year's COP28 conference in UAE. A Cooling Stocktake report will examine new prospects, review country measures already taken towards sustainable cooling, and make suggestions for political action that can cut GHG emissions from the cooling industry (Press Release 2023). Similarly, India has numerous measures in place to combat the growing threat of heat. The HAPs are India's policy responses to take measures to protect its population. However, there are numerous challenges in its implementation.

Leadership and political will to invest in resilience against the impacts of excessive heat is not uniform across levels of government. Governancearrangements at the local level to tackle heat waves are largely mobilized through efforts of centralized agencies or international NGOs/development partners. The accountability for implementation and coordination of measures included in the HAP is distributed between the state (provincial) disaster management authorities, local governments, and the district health authorities. Urban planning and implementation capacity in India is largely restricted to land use, and therefore greater attention to urban governance is also critical for developing resilience to urban heating and, likely, other impacts of climate change. There are also significant gaps in terms of developing a learning ecosystem around health impacts or the impact of interventions, including studies on cost-effectiveness and implementation research. Data systems are not yet developed enough to provide contextual information that is well integrated and in real-time. The capacity for synthesizing and applying existing information to ongoing, routine decision-making is also very limited.

Given the scale of the problem, much more needs to be done in India and elsewhere, to address the heat-related health impacts. Sustained prioritization, systematic planning, and intersectoral coordination are necessary to increase the opportunities to adapt and sustain populations.

References

Abdulali S, Abdulali L. Victims of the weather: how climate change is creating more refugees than other conflicts – Forbes India Blogs. Forbes; 2021. https://www.forbesindia.com/blog/climate-change/victims-of-the-weather-how-climate-change-is-creating-more-refugees-than-other-conflicts/

Akoijam BS, Kerketta AS, Borhade A, Khan Z, Sengupta P, Sudhakar G, Agarwal M, Khanna A, Geddam JB, Lal DK, Babu BV, Marimuthu P, Sivakami M, Kusuma YS. Living conditions of internal labour migrants: a nationwide study in 13 Indian cities. Int J Migr Border Stud. 2017;3(4):328–51. https://doi.org/10.1504/IJMBS.2017.10006517.

Balasubramanian D. Urban heat islands in India. The Hindu. 2020; https://www.thehindu.com/sci-tech/science/urban-heat-islands-in-india/article30830560.ece

Benz SA, Burney JA. Widespread race and class disparities in surface urban heat extremes across the United States. Earth's Futures. 2021;9(7):e2021EF002016. https://doi.org/10.1029/2021EF002016.

BMC. Home – Mumbai climate action plan. n.d.. Retrieved June 25, 2023, from https://mcap.mcgm.gov.in/

C2ES. Heat waves and climate change – center for climate and energy solutionscenter for climate and energy solutions. Center for Climate and Energy Solutions; n.d.. Retrieved June 25, 2023, from https://www.c2es.org/content/heat-waves-and-climate-change/

Carbonari A, Naticchia B, D'Orazio M. Buildings are a hidden source of indian cities' extreme heat. In: Eco-efficient materials for mitigating building cooling needs: design, properties and applications; 2019. p. 215–40. https://doi.org/10.1016/B978-1-78242-380-5.00008-X.

Centers for Disease Control and Prevention. Meeting report workshop environmental change and infectious disease. World Health Organization Regional Office for Europe. 2008:1–103. https://www.ecdc.europa.eu/sites/default/files/media/en/publications/Publications/0703_MER_Environmental_Change_and_Infectious_Disease.pdf

Chersich MF, Pham MD, Areal A, Haghighi MM, Manyuchi A, Swift CP, Wernecke B, Robinson M, Hetem R, Boeckmann M, Hajat S. Associations between high temperatures in pregnancy and risk of preterm birth, low birth weight, and stillbirths: systematic review and meta-analysis. BMJ. 2020;371:m3811. https://doi.org/10.1136/BMJ.M3811.

Chersich MF, Scorgie F, Filippi V, Luchters S, Huggett A, Sibanda E, Parker C, Lakhoo D, Maimela G, Rees H, Solarin I, Harden L, Hetem R, Mavhu W. Increasing global temperatures threaten gains in maternal and newborn health in Africa: a review of impacts and an adaptation framework. Int J Gynaecol Obstet. 2023;160(2):421–9. https://doi.org/10.1002/IJGO.14381.

Clement V, Kumari Rigaud K, de Sherbinin A, Jones B, Adamo S, Schewe J, Sadiq N, Shabahat Groundswell E. Groundswell Part 2: acting on internal climate migration. Washington, DC: World Bank; 2021. http://hdl.handle.net/10986/36248

Climate and Development Knowledge Network (CDKN). Putting the heat action plan into practice. NRDC; 2014. p. 1–8.

Dahl K, Licker R, Abatzoglou JT, Declet-Barreto J. Increased frequency of and population exposure to extreme heat index days in the United States during the 21st century. Environ Res Commun. 2019;1(7):075002. https://doi.org/10.1088/2515-7620/AB27CF.

Dalugoda Y, Kuppa J, Phung H, Rutherford S, Phung D. Effect of elevated ambient temperature on maternal, foetal, and neonatal outcomes: a scoping review. Int J Environ Res Public Health. 2022;19(3):1771. https://doi.org/10.3390/IJERPH19031771.

Ebi KL, Capon A, Berry P, Broderick C, de Dear R, Havenith G, Honda Y, Kovats RS, Ma W, Malik A, Morris NB, Nybo L, Seneviratne SI, Vanos J, Jay O. Hot weather and heat extremes: health risks. Lancet (London, England). 2021;398(10301):698–708. https://doi.org/10.1016/S0140-6736(21)01208-3.

EPA. Learn about heat islands. United States Environmental Protection Agency; 2022. https://www.epa.gov/heatislands/learn-about-heat-islands

Gregory J, Azarijafari H, Logan A. Urban Heat Islands. MIT Climate Portal; n.d.. Retrieved June 25, 2023, from https://climate.mit.edu/explainers/urban-heat-islands

Hari V, Dharmasthala S, Koppa A, Karmakar S, Kumar R. Climate hazards are threatening vulnerable migrants in Indian megacities. Nature Climate Change 2021 11:8. 2021;11(8):636–8. https://doi.org/10.1038/s41558-021-01105-7.

IDMC. Global Report on Internal Displacement 2023. 2023. https://www.internal-displacement.org/sites/default/files/2023-05/IDMC_GRID_2023_Global_Report_on_Internal_Displacement_HQ.pdf

IEA. The future of cooling opportunities for energy-efficient air conditioning together secure sustainable. International Energy Agency; 2018. www.iea.org/t&c/

IPCC. Working Group III contribution to the sixth assessment report of the intergovernmental panel on climate change mitigation of climate change. Intergovernmental Panel on Climate Change; 2022. p. 1–2042. www.ipcc.ch

Jairaj B, Malaviya S. Buildings are a hidden source of Indian cities' extreme heat | World Resources Institute. World Resource Institute India; 2019. https://www.wri.org/insights/buildings-are-hidden-source-indian-cities-extreme-heat

K V. The white roofs cooling women's homes in Indian slums – BBC Future. BBC. 2023; https://www.bbc.com/future/article/20230628-the-white-roofs-cooling-womens-homes-in-indian-slums

Keith L, Meerow S, Hondula DM, Turner VK, Arnott JC. Deploy heat officers, policies and metrics. Nature. 2021;598(7879):29–31. https://doi.org/10.1038/d41586-021-02677-2.

Kjellstrom T, Briggs D, Freyberg C, Lemke B, Otto M, Hyatt O. Heat, human performance, and occupational health: a key issue for the assessment of global climate change impacts. Annu Rev Public Health. 2016;37:97–112. https://doi.org/10.1146/ANNUREV-PUBLHEALTH-032315-021740.

Kjellstrom T, Freyberg C, Lemke B, Otto M, Briggs D. Estimating population heat exposure and impacts on working people in conjunction with climate change. Int J Biometeorol. 2018;62(3):291–306. https://doi.org/10.1007/S00484-017-1407-0.

Kjellstrom T, Maître N, Saget C, Otto M, Karimova T. Working on a WARMER planet. 2019. https://www.ilo.org/wcmsp5/groups/public/%2D%2D-dgreports/%2D%2D-dcomm/%2D%2D-publ/documents/publication/wcms_712011.pdf#:~:text=Excessive heat during work creates occupational health risks%3B, per cent of his or her work capacity.

Knowlton K. Climate change and health: extreme heat. NRDC; n.d.. Retrieved June 22, 2023, from https://www.nrdc.org/resources/climate-change-and-health-extreme-heat#/map%23fnh

Koll R. Death by degrees. Frontline. 2023:1–6. https://www.scribd.com/document/651970620/DeathByDegrees-RoxyKoll-Frontline-June2023#

Liu J, Varghese BM, Hansen A, Xiang J, Zhang Y, Dear K, Gourley M, Driscoll T, Morgan G, Capon A, Bi P. Is there an association between hot weather and poor mental health outcomes? A systematic review and meta-analysis. Environ Int. 2021;153:106533. https://doi.org/10.1016/J.ENVINT.2021.106533.

Madan P. As planet warms, india develops first national cooling plan. NRDC; 2018. https://www.nrdc.org/bio/anjali-jaiswal/planet-warms-india-develops-first-national-cooling-plan

Madan A. Understanding the implications and mitigation strategies for heatwaves in India. Observer Research Foundation; 2023. https://www.orfonline.org/expert-speak/understanding-the-implications-and-mitigation-strategies-for-heatwaves-in-india/

Madan P, Korsh J. Cool roofs: community-led initiatives in four Indian cities. NRDC; 2021. https://www.nrdc.org/bio/anjali-jaiswal/cool-roofs-community-led-initiatives-four-indian-cities

Madan P, Limaye V, Kwatra S, Goldstein DB. Telangana announces a groundbreaking cool roof policy. NRDC; 2023. https://www.nrdc.org/bio/prima-madan/telangana-announces-groundbreaking-cool-roof-policy

Miller S, Chua K, Coggins J, Mohtadi H. Heat waves, climate change, and economic output. J Eur Econ Assoc. 2021;19(5):2658–94. https://doi.org/10.1093/jeea/jvab009.

MODIS. MODIS web. NASA; n.d.. Retrieved September 22, 2023, from https://modis.gsfc.nasa.gov/data/dataprod/mod11.php

Mohajerani A, Bakaric J, Jeffrey-Bailey T. The urban heat Island effect, its causes, and mitigation, with reference to the thermal properties of asphalt concrete. J Environ Manag. 2017;197:522–38. https://doi.org/10.1016/J.JENVMAN.2017.03.095.

Nath S, Singh S, Swaraj S, Srinivasan V. Pathways for building more climate resilient Indian cities. pdf. Centre for Social and Environmental Innovation; 2021. https://drive.google.com/file/d/1TrIyv1gVOIAbhPpSfQuPybeBlhLDIIG2/view

National Disaster Management Authority, & Ministry of Home Affairs, G. of I. Beating the heat- how India successfully reduced mortality due to heat waves. NDMA; n.d. p. 2–15. Retrieved September 8, 2023, from https://ndma.gov.in/sites/default/files/IEC/Booklets/HeatWaveA5BOOKFinal.pdf

National Disaster Management Authority, Ministry of Home Affairs, & Government of India. National guidelines for preparation of action plan- prevention and management of heat waveof. National Disaster Management Authority; 2019. p. 1–62. https://nidm.gov.in/PDF/pubs/NDMA/27.pdf

NIH. Temperature-related death and illness – climate change and human health. National Institute of Environmental Health Sciences; 2018. p. 16–8. https://www.niehs.nih.gov/research/programs/climatechange/health_impacts/heat/index.cfm#footnote2

Norford, L., & Gribkoff, E. Heating and cooling | MIT Climate Portal. n.d.. Retrieved June 25, 2023, from https://climate.mit.edu/explainers/heating-and-cooling

NRDC. Expanding heat resilience across India: heat action plan highlights 2022- NRDC April Issue. Natural Resources Defense Council; 2022. p. 1–14.

NRDC. Expert committee recommendations for a heat action plan based on the ahmedabad experience. Natural Resources Defense Council; n.d. p. 1–19. Retrieved June 25, 2023, from http://www.censusindia.gov.in/2011-prov-results/prov_data_products_gujarat.html

Pachauri RK, Meyer L, Hallegatte France S, World Bank, Hegerl G, Brinkman S, van Kesteren L, Leprince-Ringuet N, van Boxmeer F. Climate change 2014 synthesis report. In: Kristin Seyboth (USA). Gian-Kasper Plattner; 2014. http://www.ipcc.ch.

Patel A. Indian cities can ease the impact of climate change on migrants with better affordable housing. Scroll; 2021. https://scroll.in/magazine/1007459/how-can-indian-cities-shield-vulnerable-migrants-from-climate-change-with-better-affordable-housing

PIB Delhi. Centre writes to Chief Secretaires /Administrators of all States/UTs on measures for effective management and mitigation of adverse effects of extreme hot weather on workers and labourers. Ministry of Labour & Employment; 2023. https://pib.gov.in/PressReleaseIframePage.aspx?PRID=1917571

Pillai AV, Dalal T. How is India adapting to Heatwaves?: an assessment of heat action plans with insights for transformative climate action. CPR; 2023. https://cprindia.org/briefsreports/how-is-india-adapting-to-heatwaves-an-assessment-of-heat-action-plans-with-insights-for-transformative-climate-action/

Plumer B, Shao E. Heat records broken across earth. The New York Times. 2023; https://www.nytimes.com/2023/07/06/climate/climate-change-record-heat.html

Press Release. Partners announce new ambition on sustainable cooling for COP28. UN Environment Programme; 2023. https://www.unep.org/news-and-stories/press-release/partners-announce-new-ambition-sustainable-cooling-cop28

PTI. India may lose productivity worth 34 mn jobs in 2030 due to global warming. Business Standard. 2019; https://www.business-standard.com/article/pti-stories/india-could-face-productivity-loss-equivalent-to-34-million-jobs-in-2030-due-to-global-warming-un-119070200233_1.html

Rathore V. Climate change in India: why a bus of climate migrants travels from Odisha to Kerala every week. Scroll; 2023. https://scroll.in/article/1048724/from-odisha-to-kerala-a-bus-of-climate-migrants

Rocklöv J, Dubrow R. Climate change: an enduring challenge for vector-borne disease prevention and control. Nat Immunol. 2020;21(6):479–83. https://doi.org/10.1038/S41590-020-0692-7.

Samuels L, Nakstad B, Roos N, Bonell A, Chersich M, Havenith G, Luchters S, Day LT, Hirst JE, Singh T, Elliott-Sale K, Hetem R, Part C, Sawry S, Le Roux J, Kovats S. Physiological mechanisms of the impact of heat during pregnancy and the clinical implications: review of the evidence from an expert group meeting. Int J Biometeorol. 2022;66(8):1513. https://doi.org/10.1007/S00484-022-02301-6.

Sawant NH, Sanjeev A. Climate refugees in India: a wake-up call for an inclusive policy. India Q A J Int Affairs. 2022;78(2):371–80. https://doi.org/10.1177/09749284221090706.

Somerville R, Trenberth K, Meehl J, Masters J. Heat waves and climate change. Climate Communications Science & OUtrecha; 2012. p. 1–14.

Tasgaonkar P, Zade D, Ehsan S, Gorti G, Mamnun N, Siderius C, Singh T. Indoor heat measurement data from low-income households in rural and urban South Asia. Sci Data. 2022;9(1):1–11. https://doi.org/10.1038/s41597-022-01314-5.

UNEP, & IEA. Cooling emissions and policy synthesis report: benefits of cooling efficiency and the Kigali Amendment Homes workplace medicine institutions transport data centers food cooling emissions and policy synthesis report. 2020. www.unep.org

Venugopal V, Shanmugam R, Kamalakkannan LP. Heat-health vulnerabilities in the climate change context–comparing risk profiles between indoor and outdoor workers in developing country settings. Environ Res Lett. 2021;16(8):1–14. https://doi.org/10.1088/1748-9326/AC1469.

Walker JT. The influence of climate change on waterborne disease and Legionella: a review. Perspect Public Health. 2018;138(5):282–6. https://doi.org/10.1177/1757913918791198.

WHO. Heat and Health. World Health Organization; 2018. https://www.who.int/news-room/fact-sheets/detail/climate-change-heat-and-health

World Resources Institute India, & C40 Cities Network. Mumbai sets a mitigation goal to achieve net-zero by 2050. India Water Portal; 2022. https://www.indiawaterportal.org/articles/mumbai-sets-mitigation-goal-achieve-net-zero-2050

Inhabiting an Aqueous World: Design Research for the Environment and Communities

9

Deepta Sateesh

Abstract

The monsoon terrain of the Southwestern Ghats in India has been territorialized through an act of colonial power, which either settled or marginalized peoples, practices, ecologies, and particularly, the weather—privileging a wet-dry binary and spatializing a mobile wet world. This chapter focuses on indigenous "wet ontologies" that privilege everyday practices across time, continually making a home in rain and wetness. The rain and multiple conditions of wetness across seasons are undeniable and inescapable, yet policy frames do not recognize them as ordinary, as a medium through which life occurs and is ever-present. This chapter addresses this concern by suggesting the possibility of new policy frameworks designed from lived experiences, responding to flux and wet ontologies. As we walk through an inherited past, a conflicted present, and possibilities for the future, the chapter explores the realities of soaking ecologies. It draws upon insights from primary research conducted by walking with inhabitants of the Ghats in Wayanad, Kerala.

Keywords

Aqueous world · Communities · Wetness · Southwestern Ghats · SDGs · Wayanad · Deluge

D. Sateesh (✉)
HVDC Control and Protection Systems Engineer, General Electric, Stafford, UK
e-mail: deepta.sateesh@manipal.edu

H. Brand et al. (eds.), *Climate Change and Public Health*,
https://doi.org/10.1007/978-981-95-0895-2_9

9.1 Introduction

The monsoon terrain of the Southwestern Ghats is territorialized through an act of colonial power that either settled or marginalized peoples, practices, ecologies and particularly, the weather, privileging a wet-dry binary and spatializing a mobile wet world. These coastal mountains have been inhabited by a multitude of inhabitants, human and non-human, who understand dwelling in a monsoon terrain by moving, occupying, and temporally appropriating dynamic conditions of "wetness" in their ordinary everyday lives (Sateesh 2020a). This chapter focuses on indigenous "wet ontologies" (Steinberg and Peters 2015) that privilege everyday practices across time that continually make a home in rain and wetness.

Today, these ordinary practices remain outside the inherited colonial framework, their understanding, and the ways of living in a dynamic wet ground excluded from dominant narratives and knowledge systems. The gap between dwelling experience and climate policy is dramatically wide, particularly in recent years when the terrain has experienced disasters attributed to rain, and efforts to mitigate these disasters further harden and fix this mobile world. Climate policy requires a different approach for this place fed by the monsoonal rains, as the language it is written in is far removed from lived realities. The rain and multiple conditions of wetness across seasons are undeniable and inescapable, yet policy frames do not see them as ordinary, as a medium through which life occurs, ever-present. This chapter responds to this concern, by suggesting the possibility of new kinds of policy frames designed from lived experiences that respond to flux and wet ontologies as we walk through an inherited past, a conflicted present, and possibilities for futures, by dwelling in the realities of soaking ecologies (Bhattacharyya 2018). The Southwestern Ghats is a world dominated by wetness, movement of materials and inhabitants, a wet, humid, rainy, boggy world (Ingold 2010). Here, the focus is on the significance of the role art and design have played in constructing not just the dominant representations of places in the Ghats but also created conflicts, between the outsider and inhabitant, and atmosphere and earth.

9.2 A Shifting Landscape

During the past few years, the Ghats and other regions have experienced dramatic overflows of waters and shifting earth—what we term landslides and floods, along the western coast and mountains. The devastation to the landscape and livelihoods, on August 9, 2018, when waters from the largest dams in Kerala, Idukki, and Idamalayar were opened for fear of the dams breaking under the pressure of the amount of water that had gathered in the reservoirs. The officials had waited far too long to release water and, rather than a gradual release, had to open multiple sluice gates simultaneously. Madhav Gadgil was on the news sharing that the devastation was caused by thoughtless construction projects (for development) and the lack of clearly demarcating ecologically sensitive zones for conservation in the Ghats. Most of the areas that were affected by the events were colonial monoculture plantations, teak forests, tea and coffee plantations, and intensely cultivated swales, called *bailu* (Kannada) and *vayal* (Malayalam), throughout Wayanad, Coorg, and South and North Canara. The plains and coast saw waters everywhere, the backwaters and canals gushing, while many of the locals near the coast took to their houseboats during this time of inundation. Roads were torn and washed away, along with buildings, and expansive low-lying cultivation areas inundated. After the floods of 2018, the inhabitants of the affected areas experienced many health issues, including skin, respiratory, and bacterial infections from stagnating pools.

These floods and landslides have been largely attributed to climate change, of when, where, and how much rain has fallen—that a certain amount of rain that had fallen on those few days was more than usual in a particular geography close to the dams—termed as a "natural disaster" highlighting rain as the culprit. However, the floods and landslides of 2018 and 2019 are not necessarily markers of only increased rainfall. Following Gadgil's clues, perhaps we can also attribute it to the way humans have attempted to control and manage land and water as separate from each other, a misunderstanding and misrepresentation of the Ghats terrain, and a lack of adequate understanding of the weather, the way the earth transforms with it, and the abilities to respond in this dynamic world. It is not possible to predict exactly where, when, and how much rain will precipitate. And when it precipitates, it becomes water that gets soaked, muddiness, mist, and wet materials transforming them, moving through them, into depths of the atmosphere and earth.

In the current discourse around climate change is the Anthropocene, a geological time that attributes blame on humankind, for intensified and extractive human activities interrupting natural processes of ecologies and planet. At this moment, perhaps humans need to take credit (blame?) for; in this scenario, these are design disasters that stem from fixing space and regularizing time. The Idukki and Idamalayar dams, like all the other dams in the Ghats, are built for power generation and need expansive reservoirs of water to be held for months, fed by redirected streams and rivers, channels, and pipes, making other places parched dry and wetting soils that have little capability to hold water. They demand permanent infrastructures privileging land and taking up a fixed space/area, and still waters that a multitude of inhabitants, human and non-human used to depend on for drinking, cleaning, and other daily practices—which they have no access to now. These waters are redirected to serve cities in the distance with electricity. What is instead desired is activity and movement in and of waters and wateriness for the element or medium's health. Consideration needs to be given to the monsoon season in the Ghats and beyond, as it is a time of great change, renewal, revitalization, initiation, growth, decay, motion (with the movement of water, inhabitants, materials, and atmosphere) and the emergence of new things, processes, and practices.

9.3 Water or Wateriness?

At the intersection of climate change and public health lies the concern for the way we conceive and perceive water. We experience life in inescapable wateriness, a gradient of events such as rain, mist, humidity, and conditions of wetness, waters being in plants, soils, humans, and the atmosphere. But we learn about and imagine water as something separate from land and atmosphere and color it blue. This is a view from above, an inherited way of seeing the world through colonial eyes. In particular, the Western Ghats has, over the past 250 years, been viewed, imaged, and territorialized through an act of surveying and representation through the act of making maps and other drawings that attempted to settle peoples, practices, ecologies and marginalize the changing weather, privileging a dry imagination that fixes and spatializes this mobile world.

This is a separation of water from land and atmosphere and emerges from colonial thinking about places—that water and wateriness lie outside of human habitat. A "wet ontology"[1] (Steinberg and Peters 2015) considers a ubiquitous condition of wetness of our world, particularly the Southwestern Ghats of India, with waters seeping, oozing, soaking, evaporating, precipitating, and transforming. This is a worldview of inhabiting an aqueous world that privileges everyday practices across time that continually make this terrain, in rain and with wetness, such as growing food and making shelters. The inherited colonial framework gives no voice to this wetness, excluding local communities' understandings from dominant narratives and knowledge systems, and is subsequently left out of planning and development, laws, and policies. The act of recording data and drawing/representing information that appears as static visuals disallows change to be accommodated into a broader more inclusive understanding of the Ghats through time. After separating water from land, roads, industries, and high-density housing are built, and we design permanent infrastructures to manage and control water, like dams, levies, drains, and such. This separation of water from land has a direct impact on the health of the environment and its human and more-than-human inhabitants, as it tends to control and manage natural processes, to solely serve infrastructural needs.

Interestingly, situated practices are rooted in a particular understanding of dynamic shifting ecological processes and the ongoing reappropriation (by local communities through natural processes) of terrains and materials that constitute them, across time in the Ghats, appropriated for dwelling in a terrain that requires continual care for sustaining the health of ecologies and inhabitants (da Cunha 2019). Engaging in the Ghats requires a different imagination through everyday ordinary practices and is temporal and more complex than the rationality of the dominant conceptualization (of land separate from water, or a wet-dry binary colonial worldview embedded and institutionalized through laws and policies) that has designed landscapes for production and protection. This other imagination, of inhabiting an aqueous world, privileges a gradient of wetness as ordinary normal conditions of the world experienced on foot.

[1] Steinberg and Peters describe "wet ontology" as a way of immersing in and understanding the ways of waters in oceans—that water and watery beings are voluminous, dynamic, and permeate other materials and processes, transforming them through movement, rhythms of time, and practices, and beyond planetary forces, considering water as a primary medium of the world (2015).

9.4 Dissonances and Conflicts

In the Southwestern Ghats, and particularly since its designation as a UNESCO World Heritage Site in 2012, conflicts have escalated on the ground, emphasizing the need to reveal a more fundamental source of these clashes. This environment articulated through maps and other drawings is designed to become a static habitat for humans by imagining and drawing lines on paper. However, these are primarily ways to abstract space and freeze time, stilling the world now convenient to draw plans, sections, maps, accounts, and calendars, all universally regularizing and making predictable this dynamic environment of the Ghats, and deem them true representations of the world of the past and the world of the future. To think more inclusively about the monsoonal region of the Ghats, it is important to imagine it as a world of a gradient of wetness from least wet to most wet. Towards this are two possible conceptualizations of the world one of the ideas of a static surface of land that separates water from it and separates atmosphere and earth, wet from dry, a function of extracting water out of material making it devoid of wetness, and a land water divide permanent (da Cunha 2019). Another is a world of processes of seepage and soaking, of holding and releasing moisture, of inhaling and exhaling, transforming over time with water (rain) becoming multiple conditions of wetness (from least wet to most wet). These two worlds that clash in the Ghats are between the voyeur and the walker (Certeau and Michel. 1984, 91–93). The voyeur, or the colonial administrator, surveyor, or scientist, gazes, analyzes, freezes, or arrests the world in space and time through the visual (eye), using only a single sense, devoid of a sensibility of being in the world. The voyeur writes to read or reads to write, to represent the world in a way that can capture/record, interpret, and re-present to manage, strategize, assert power, and control (Edney 1997). As one moves the eye from being in the ground, up and above the ground, the world appears as only a static image to the eye, with other senses becoming inactive, and from afar the world moves less as one is unable to see the nuances in changes that occur at a scale of human engagement on the ground. The view from the above is advantageous, although it is only a partial reality (Haraway 1988), a privileged perspective as it hides other aspects of reality, of change, movements, relationalities, and time, what James Scott terms as the "optics of power" (Scott 1998, 253). The visualization has been constructed to separate nature from culture, outside from inside, and water from land, creating conflicts and violence.

This quasi-scientific lens is one of a world that is wholly knowable, stable, and predictable (Sarukkai 2012, p. 23–24), one of which we draw lines to organize and transmit and manage and control space, materials we call resources, inhabitants, time, articulating the world as such to understand and know it. However, there are things of the world for which these kinds of lines cannot be drawn, and these notions remain outside of the mechanistic view of the world, such as material movements, other sentient beings, and the conditions of wetness that shape-shift multiple grounds for dwelling in this monsoon terrain. There is a world outside of the colonial eye, an imagination that exists in the everyday contemporary that has not been engaged in thinking about new futures (that may extend these locally situated practices), by immersing in and inhabiting a world of wetness.

Being in the world is being in the ground of the Southwestern Ghats, with all senses and sensibilities actively making the world, an immersion into context and to be context-sensitive (Ramanujan 2004). But when seeing from above, there are no sensations but vision. And from up there, the world is still a "gigantic mass [is] immobilized before the eyes" (Certeau and Michel. 1984). This is the world that is familiar to us through an act of familiarizing our formal education system and the disciplines separated from one another. The conflicts are at a fundamental level of an inadequate understanding of the complexities of the environment and its ordinary processes. When we inhabit wetness by walking, how can we see it as local, in a "rainy terrain" (Cunha and Dilip. 2018)? The architect, planner, scientist, and engineer must shed their view from above, one that is taken for granted to be real and true and see what is here and now on the ground in the environment, through wetness (of dynamic material conditions), and what is changing.

Speculatively, one can experiment with these notions, to develop them as practical frames towards new futures of nature-culture synchronicities (something that the voyeur is unfamiliar with). We engage with the seemingly unknown to speculate the diverse possibilities, by entering the world of the walker. Here are some understandings through primary research of walking with inhabitants of the Ghats, in Wayanad, Kerala. What is distinct about some farming families in this region is their understanding of different kinds of precipitation and rain (Viju 2012) and how to correspond and adapt to them (Figaredo n.d.), each initiating diverse practices of not just cultivating the earth, but growing foods such as pepper and vanilla that require particular conditions of atmosphere and tree trunks, of moisture and humidity, to manifest (Shaji 2019). These conditions are brought on by seasonal changes in weather experienced in the locale or context—such as *noolmazha*, thread-like precipitation that has become less common (Viju 2012), and mist required for seed conservation, and *karakadam* the time of heavy rain and thunderstorms (Sateesh et al. 2018). This is a different understanding of time (at least one other) from the abstract, predictable, reliable, and accurately measured time of the Gregorian calendar; this is an understanding of time that is of the wet world of the Western Ghats, of diurnal flux and changes in seasons and weather systems, which has not been paid attention to in planning and design (of materials, infrastructures, and policy) but is significant as it rains for more than half the year. The lifeworld of the Ghats is continually changing in minute and dramatic ways, of operations of growth, movement, and soakings. Here time is embedded in the everyday ordinary, with the movement of materials and inhabitants (human and non-human).

Although many farms and communities were unable to cope with the devastation the floods and landslides in 2018 brought, there were others whose creative practices passed down from generation, working with rain, precipitation, inundation, and the changing atmosphere and earth. These practices include the conservation of diverse seeds, working with the slopes between low and high grounds to create nurseries and transplanting, using medicinal concoctions of wild herbs to enrich the soils and saplings, and planting fruit trees amidst the paddy fields that attracted birds whose droppings strengthen roots. Recorded everyday practices (through the author's primary research) of engaged understanding of this seeping world include curing seeds in misty conditions; thatching shelters made of mud; plastering porous walls with a mix of grass, mud, cow dung, and ash; planting diverse crops along slopes also ensures diversity of soils, some that do not wash away easily and that some varieties of paddy, after a deluge, may not be consumable by humans but is so for other creatures who then fertilize the earth, and siting home in a way to privilege soakings across dwelling practices, as well as understanding decay as a material change. In this way, these small farms transform seasonally, each ground of different rhythms synchronized with changing conditions, from bearing produce to laying fallow, of rhythms of movements and pauses of inhabitants and airs. The inhabitants of Wayanad may be creative practitioners, continually constructing a "dynamic landscape matrix" (Perfecto et al. 2009), of multiple grounds and practices that nurture and sustain these practices, not just for humans but include the participation of other sentient beings as well as diverse materials and movements (Sateesh 2020b). This terrain is a complex shifting matrix, a gradient of wetness with non-human and human inhabitants moving with it, and through it, oscillating in between, ever becoming, and inhabiting places through ordinary everyday practices, moving through moments and inhabiting all, at once.

How do they do this? How do they read a different world? Is it possible for us (the urban, educated dwellers) to at the very least partially convert into this sensitive, inclusive, and responsive inhabitant? How do they know how, where, and when to site certain practices? In Tulunadu and other places in the southern peninsula of India, communities read the skies for "rain" stars that indicate types of precipitation and what to cultivate, when, and how (Amulya 2017). Some suspect these local people's knowledge is superstitious, not considering diverse forms of transmission—such as songs (*paddanas*), material practices of growing food and homemaking, myths, and performances (*bhuta kola* and *nema*). Some of these forms and narratives change over time and through generations, incorporating a changing world.

9.4.1 Traversing the Ghats

Wetness and rain are inescapable in the Ghats. Local inhabitants dwell in a world that is always some condition of wetness, an aqueous world, an ecotone[2] (Neimanis 2012), or a threshold that they correspond with. These inhabitants can read change from being dwellers in this world, negotiating by walking, a fundamental movement to carry on with life; it is an ordinary everyday practice through a ground or milieu (Ingold 2010) that is devoid of subjectivity—it is as well as universal, a real way of being immersed. When we walk, how do we see and what do we see? By walking through a ground—ground is like an eco-tone—we traverse through the atmosphere and earth and encounter inhabitants, rhythmically responding and shaping and moving with and through. This walking practice, or practice of movement, is a

[2] "Ecotone" is a place of transition between two or more ecologies, a place of dynamic change and transformation, such as estuaries and swamps (Neimanis 2012).

meta-practice, one that holds all other practices and goings-on. Henry David Thoreau elaborates on walking as a way of wandering, of giving in to the practice of walking without thinking about it but thinking through it. Walking, to Thoreau, is a way of getting lost, not in a manner that one is clueless about one's whereabouts, but in a manner to find one's way through the environment, to be a creative practitioner in the everyday. This "getting lost" is a type of wandering that is of being immersed in the context (Thoreau 1862) and of a sensibility that acknowledges that phenomena are ephemeral, never to occur in the same way again—a world unpredictable. Tim Ingold further explores walking as being a material practice that involves paying attention (Ingold 2000) or demonstrating care through practice. This being-in-the-weather world is possibly an existence to carry out practices, operations, activities, and makings.

This is also about the ordinary rather than the exotic—practices such as walking, thatching, sitting, searching for water, foraging, harvesting, or mending fishing nets, peeling/shelling *adike*, finding certain medicinal plants, and so on. This is the act of *dwelling* (Heidegger 1971), of world-making. It is not just one practice, nor space, that defines the makings of the world, but a whole array of practices, skills, and movements that enable being in the world—a multiplicity (Heidegger 1996) and entanglement in meshwork (Ingold 2011) of doings. In Wayanad, this multiplicity is held in material transformations and practices—paddy is grown, harvested, thrashed, the grasses cured, becoming thatching for roofs, and then decay into the fields becoming earth—all corresponding with conditions of wetness, with time. These communities dwell in watery soaking ecologies, following conditions of wetness for daily life. Each practice is intertwined and interdependent with other practices.

How do we shift from the dominant to the world of wetness, and how does this affect practices of design, environmental health, and planning? If we are now privileging wetness, rather than resisting it, learning to work with changing patterns, and drawing is fundamental to design, what kinds of lines do we draw now? What kinds of lines are of a gradient of wetness?

9.5 Design Research for Situated Ground Action and Localized Policy Making

To address the kinds of problems that are seemingly "natural" disasters, universal policy frames have been created by the UN to formulate country-specific policies, particularly around reducing temperatures, reducing dependency on fossil fuels, and other solutions put forth in the UN Sustainable Development Goals (SDGs) and other policy frames (*United Nations* 2015). These are solutions to defined problems, where design and technology step in to respond with new interventions. However, these "things" are responses to linearly defined problems, not the complex wicked ones they have become, perpetually creating new problems with the solutions, such as a solar farm that reduces fossil fuel dependency but disturbs biodiversity and access to water, for example. Climate policy at the global level deals with (atmospheric) temperature (as a measure and indicator) and water level change (terrestrial) rather than reading and working with weather and changing ground. It is geared to fix and stabilize climate following globally accepted norms of quality of life for inhabitants, an anthropocentric (political and academic) agenda.

The SDGs are based on universal values and built on a global platform (United Nations 2015). However, the universal principles they are based on are often lost in translation or delinked when applied, whether for a country or implemented at a local level. The importance and intended impact of the SDGs are often reduced through inaccessible policy documents, a lack of relevant and appropriate language, and a misreading of traditional knowledge and local practices through universal frameworks and visualizations of the world. Here, it is ever more evident that the dynamic context and mobile inhabitants have never been considered on their terms. This seeing the world as static excludes everything non-human, not just weather but other life forms, material processes, and changing conditions. In addition, most policies are derived from concerns of the urban cities and do not see beyond to include in the frame of thought. There are many ways communities dwell beyond the "city," such as Wayanad, Tulunad, and South Canara, for example, in which situated understandings of place, weather, and inhabitants are privileged, where communities interdependent with natural processes in their daily lives seem to be able to respond to the broader issues and philosophical underpinnings of climate change (and perhaps even help frame solutions for the "city"), through more tactical methods that deal with unpredictability and change.[3] That is, rather than look at the urban for solutions to their problems and those of other "places," it may be possible to draw responses to these issues from places and knowledge that have already been included in their situated practices, the participation of atmosphere, earth, inhabitants, and time.

The following is a case of a research project in Wayanad, with an opportunity to engage in the ground, in an immersive manner, to uncover knowledge and practices that are sensitive to the ecology, weather, and earth. Here, design research was a process of discovery enabling new structures to emerge that initiated new thoughts, practices, and policies. Design is a process of exploration, recording, and synthesizing. By inhabiting a lens of seepage (and a gradient of wetness), the research uncovered a more holistic and appropriate place-based understanding to respond to, rather than deploying top-down "universal" goals and strategies, such as the SDGs. A contradistinction between the two approaches (immersive and top-down) reveals a non-linear open system that allowed for emergent ideas and practices to be continually integrated and articulated into policy. To design a sensitive and responsive framework that rethinks the SDGs for a place, there is a need to (i) interpret the SDGs appropriately towards contextualizing them; (ii) investigate existing practices in the field that correspond with the weather, other

[3] One example is Sustainable Development Goal 13 on Climate Action which states there is an urgency to "combat" global temperature rise. This is a phenomenon of urbanized areas where the urban heat island effect with large expanses of permanent infrastructures absorbs and emits heat, something that is uncommon in rural and forested areas.

inhabitants, and earth processes, through creative practices that record change; and (iii) invent new methods of creative documentation and representation to be inclusive of inhabitants, holders of environmental knowledge. This was significant, as the inhabitants in Wayanad do not see their home environments and conditions as recorded and measured by instruments, measured numerically, and analyzed to understand trends and fluctuations. Rather they engage physically and materially, immersed in the changing conditions. The base research for this has already been undertaken (Sateesh et al. 2018) and needs further investigation in the field and synthesis of new kinds of data based on seepage, to construct tools and write policies from the ground.

In the aftermath of the 2018 deluge through Wayanad, as the district began to respond, rebuild, and formulate a new District Development Plan, two questions grounded in the lens of seepage were asked: (i) How can new tactics be initiated into a development plan that works with a mobile ground, and (ii) how can new frameworks be devised for policy, both drawing from situated practices of farmers in Wayanad? By including monsoon rains as ordinary and recurring each year, and thus the operation of seepage through materials, it became possible to reframe the way everyday cultivation and dwelling practices of inhabitants are understood, practiced, and represented, by privileging the relations between each different kind of rain with practices associated and processes initiated at that moment (rather than aligning cultivation with the Gregorian calendar). Another significant yet taken-for-granted research method is the ordinary practice of walking (Thoreau 1862; Sateesh 2020b); walking has been and continues to be a fundamental practice of traditional farming communities, traversing terrains while tending, cultivating, harvesting, and growing; it is a practice of attending to the world simultaneously gathering knowledge. The approach of inhabiting wetness while actively engaging in walking in this manner provided a unique opportunity to resist the dominant static view of the world from above, understanding an active landscape that is continually in flux (Ingold 2011), while moving through it.

From the research, one fundamental practice learned from traditional farmers in Wayanad is to sculpt the terrain to slow and hold wateriness through temporal earth constructions, as mentioned above, following the movement of water along the slopes to the *vayals* (swales). This movement and slowing cannot be visualized in plan-view, instead drawing sections of terrain over time enabled seeing change across diverse practices of growing and tending, to relate to these practices to changing seasons and rain events. The region that was most affected by landslides was Pozhuthana Panchayat, Wayanad (Fig. 9.1). These colonial tea plantations and forests slipped in the deluge, leaving behind trails of clayey mud. Here, a tool was developed to respond to the gradient of wetness through the atmosphere and earth, considering slopes and soils, maintaining heterogenous plantings, as well as inviting greater diversity of fauna across seasons—collectively engendering multiple future practices and processes that are responsive and inclusive of the earth and atmosphere (Srivastava 2019) (Fig. 9.2). From these place-based understandings through inhabiting seepage, new situated policy frames were designed that consider the dynamic

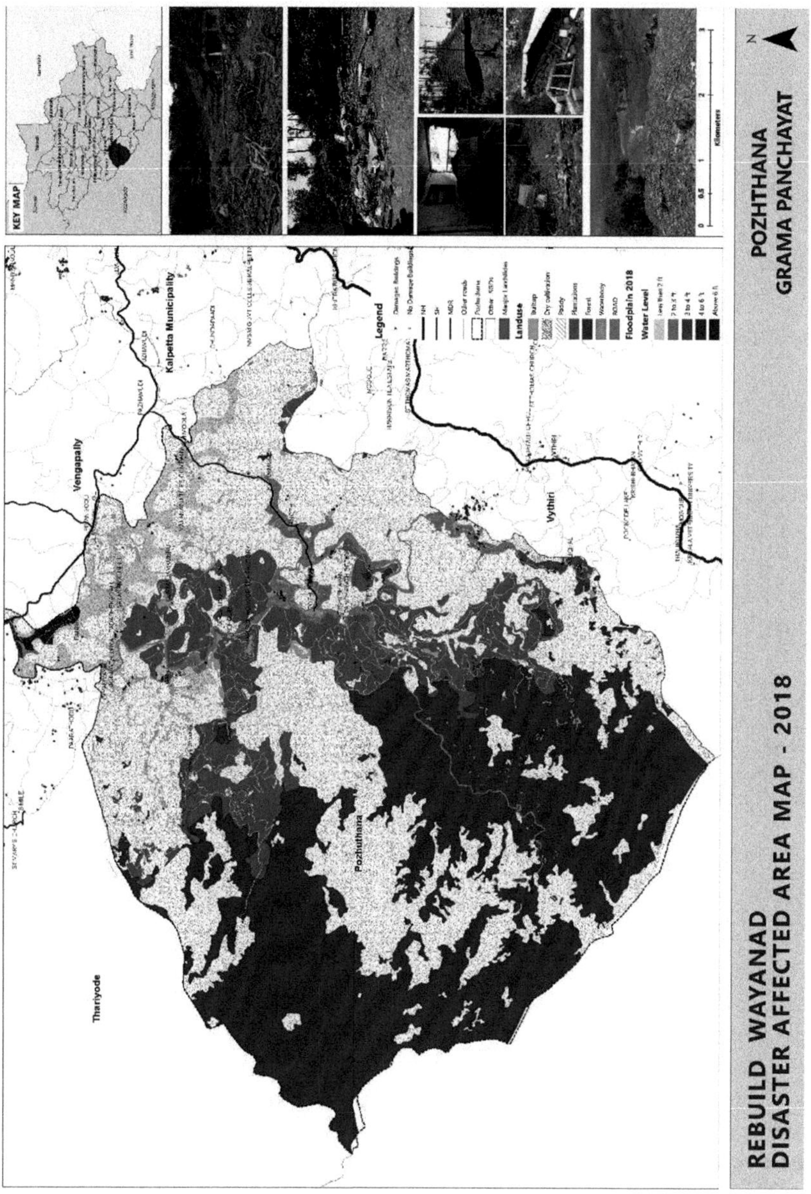

Fig. 9.1 Map of Pozhuthana Panchayat, Wayanad, showing disaster-affected areas. (Source: Falguni Gupta and DEL Laboratory, Srishti Manipal Institute of Art, Design and Technology, MAHE)

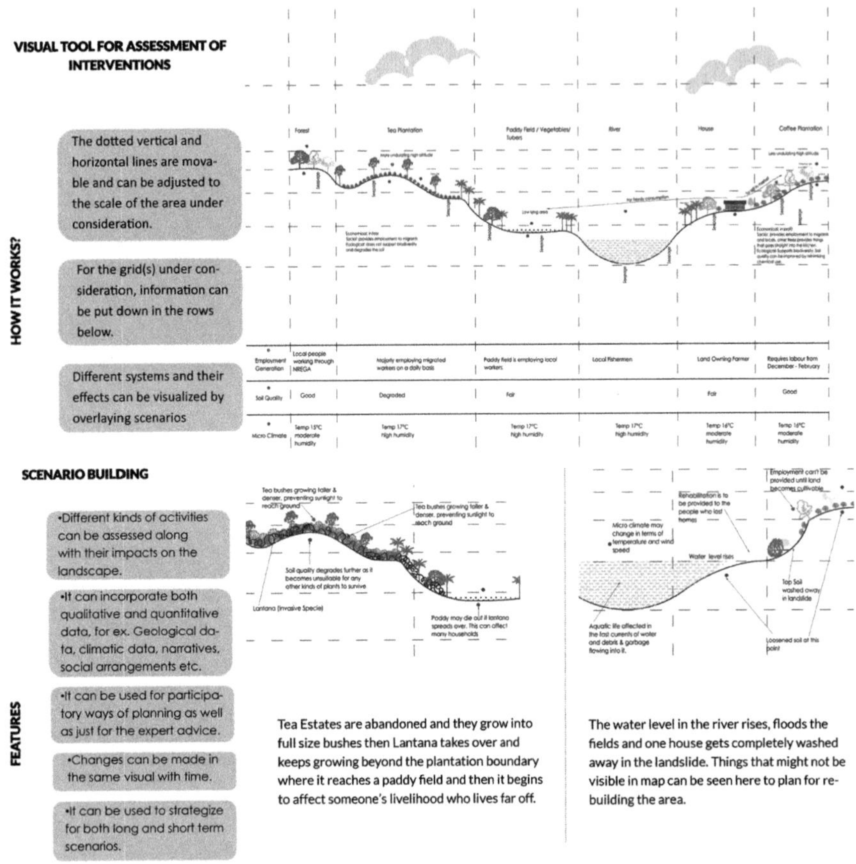

Fig. 9.2 Assessment tool designed for scenario-building in Pozhuthana Panchayat, Wayanad. The underpinning is a wet ontology. (By Srivastava 2019)

environment of the Ghats. This emergent and autochthonous framing, coming from a sectional imaging and wet imagination, offers a basis for new policies that respond to the complexities of Pozhuthana's terrain, communities, livelihoods, and more-than-human inhabitants, simultaneously critiquing the inadequacies of the language and framing of the UN Sustainable Development Goals (Gupta 2019) (Fig. 9.3). These complementary designs, one that responds to dynamic relations amongst inhabitants, earth, and livelihoods, and the other of situated policy-framing, reveal several significant ideas that privilege place-based engagements. They include the importance of (i) rain and conditions of wetness across seasons, their relations to the earth and temporal constructions; (ii) walking (being on foot) as fundamental to research; (iii) diverse ways of drawing lines that include other sentient beings and materials; and (iv) enable the weather to participate in synthesizing intricate and intimate relations in an enlivened seasonal ground.

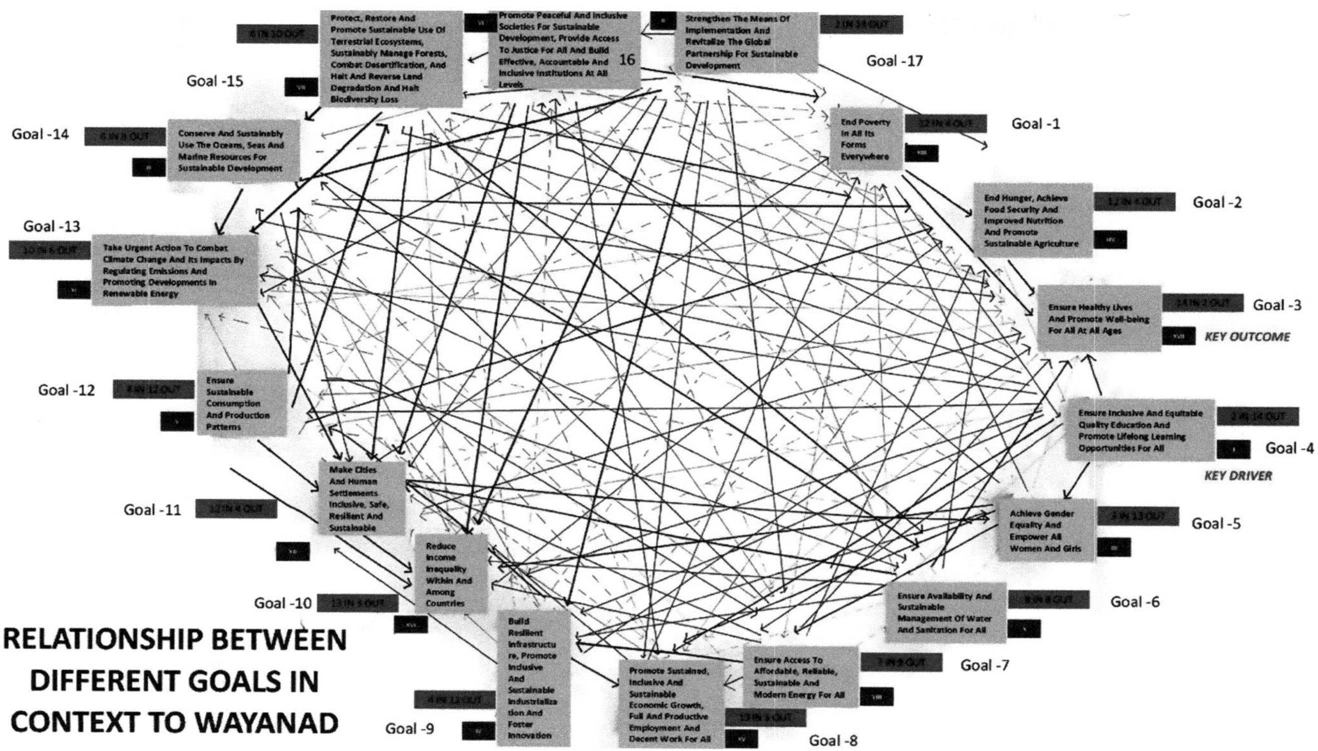

Fig. 9.3 The relationship amongst different Sustainable Development Goals in the context of Wayanad, a process of deconstructing and unlearning. (By Gupta 2019)

9.6 Conclusion: From Context-Free to Context-Sensitive

The understanding of climate and "natural" disasters can be traced to colonial understandings of place, emerging from the Enlightenment period. In Wayanad, these universal ways of seeing and settling the surface, which separates atmosphere and earth, and seeing the weather in this separation, begin to reveal ways in which lines are drawn to portray places and make sense of them to interpret (as backward) and offer interventions (to civilize). In Wayanad and the Ghats, a place where it rains at least 6 months a year and is always humid and warm, understanding the weather with dwelling and everyday practices becomes significant to understanding how to change and synchronizing with change occurs. Fundamental to this way of seeing, through seepage or a wet ontology, is the ordinary practice of walking. Plans, laws, and policies do not take into consideration these modes of engagement as they rely primarily on representations (static evidence) created through scientific engagement (objectivity and empiricism). These plans, along with international policies, created by distancing oneself from the environment based on sciences of the natural world, of a drained world, make invisible the reality of not just the gradient of wetness but also the operations of an enlivened ground that can continually soak, hold, and release moisture. Water is not H_2O. It is elemental and material. Water does not move linearly—it seeps, oozes, infiltrates, soaks, inundates, precipitates, and rises, defying gravity and moving non-linearly through materials, our bodies, and the atmosphere.

References

Amulya B. No fishing, stars that bring rain, and other legends around the breathtaking monsoons in Western Ghats. The Better India. 2017:2017. Accessed from https://www.thebetterindia.com/108306/monsoons-western-ghats-stories-beliefs-rains-nature/

Bhattacharyya D. Empire and ecology in the Bengal Delta – the making of Calcutta. New Delhi: Cambridge University Press; 2018.

Certeau D, Michel. The practice of everyday life. Berkeley: University of California Press; 1984.

Cunha D, Dilip. River literacy and the challenge of a rain terrain. In: Rao VD, editor. Critical humanities from India: contexts, issues, futures. London/New York: Routledge; 2018. p. 177–204.

Da Cunha D. The invention of Rivers: Alexander's eye and ganga's descent. Philadelphia: University of Pennsylvania Press; 2019.

Edney M. Mapping an empire – the geographical construction of British India, 1765–1843. Chicago/London: The University of Chicago Press; 1997.

Figaredo L. Climate change adaptation strategies of Wayanad tribes. Climate South Asia Network; n.d.. Retrieved from http://climatesouthasia.org/climate-change-adaptation-strategies-of-wayanad-tribes/

Gadgil M, Krishnan BJ, Ganeshaiah KN, Vijayan VS, Borges R, Sukumar R, Noronha L, Nayak VS, Subramaniam DK, Varma RV, Gautam SP, Navalgund RR, Subrahmanyam GV. Report of the Western Ghats ecology expert panel. The Ministry of Environment and Forests, Government of India; 2011.

Gupta F. Redefining sustainable development goals, Pozhuthana Panchayat, Wayanad District, Kerala. Srishti Institute of Art, Design and Technology; 2019.

Haraway D. Situated Knowledges: the science question in feminism and the privilege of partial perspective. Fem Stud. 1988;14(3):575.

Heidegger M. Building dwelling thinking. In: Poetry, language, thought. (trans: by Albert Hofstadter). New York: Harper Colophon Books; 1971.

Heidegger, M. Being and time, (trans: by Joan Stambaugh). Albany: State University of New York Press; 1996.

Ingold T. The perception of the environment – essays on livelihood, dwelling and skill. London: Routledge; 2000.

Ingold T. Footprints through the weather-world: walking, breathing, knowing. J R Anthropol Inst. 2010;16(Making Knowledge):S121–39.

Ingold T. Being alive: essays on movement, knowledge and description. London/New York: Routledge; 2011.

Neimanis A. Hydrofeminism: or, on becoming a body of water. In: Gunkel H, Nigianni C, Söderbäck F, editors. Undutiful daughters: mobilizing future concepts, bodies and subjectivities in feminist thought and practice. Palgrave Macmillan; 2012. p. 96–115.

Perfecto I, Vandermeer J, Wright A. Nature's matrix – linking agriculture, conservation and food sovereignty. London: Earthscan; 2009.

Ramanujan AK. Is there an Indian way of thinking? In: Dharwadker V, editor. The collected essays of A. K. Ramanujan. New Delhi: Oxford India Paperbacks; 2004. p. 34–51.

Sarukkai S. What is science? New Delhi: National Book Trust; 2012.

Sateesh D. Movement and place-making in a monsoon terrain. Dearq. 2020a;26:16–27. https://doi.org/10.18389/dearq26.2010.02.

Sateesh D. A walk through Malegalalli Madumagalu—lines that uncover new ontologies. Rupkatha J Interdiscip Stud Humanit. 2020b;12(1):1–15. https://doi.org/10.21659/rupkatha.v12n1.39.

Sateesh D, Gupta F, Srivastava A. Sustainable development goals and climate resilient strategies, Wayanad, Kerala – a pilot project. Bangalore; 2018. https://dellaboratory.files.wordpress.com/2019/04/1811_wayanadreport-full-web.pdf

Scott J. Seeing like a State. New Haven/London: Yale University Press; 1998.

Shaji KA. Climate change threatens Wayanad's agrarian and tourism prospects. Mongabay; 2019. Retrieved from https://india.mongabay.com/2019/10/climate-change-threatens-wayanads-agrarian-and-tourism-prospects/

Srivastava A. Inquiring land and livelihood relationship for sustainable economic development. Srishti Institute of Art, Design and Technology; 2019.

Steinberg P, Peters K. Wet ontologies, fluid spaces: giving depth to volume through oceanic thinking. Environ Plann D: Soc Space. 2015;33(2):247–64. https://doi.org/10.1068/d14148p.

Thoreau HD. Walking. The Atlantic; 1862.

United Nations. Transforming our world: the 2030 agenda for sustainable development. United Nations; 2015. Retrieved from https://sustainabledevelopment.un.org/content/documents/21252030 Agenda for Sustainable Development web.pdf

Viju B. Opposition to Gadgil panel report not in state's interest, says expert. Times of India. 2012:2012.

A HIA of Climate Change in Wales

10

Liz Green

Abstract

Health Impact Assessments (HIA), at its core, seeks to identify who is going to be affected in the population by a policy or plan and how it will impact their wider health and wellbeing. This chapter primarily focuses on an HIA conducted in Wales, which centred around the health, well-being, and equity impacts of climate change. It describes the process and results of the HIA and highlights how HIA is a useful tool for policymakers considering action on climate change. The chapter briefly outlines the quantitative aspects while also emphasizing the benefits of qualitative evidence within the process—for example, stakeholder engagement and interviews with key informants from a range of national organizations. Finally, the chapter illustrates how an HIA can help identify the differential effects of climate change on the health and well-being of diverse populations.

Keywords

Health Impact Assessment · Wales · Climate change · Stakeholder involvement · Participation · Evidence-based decision making

10.1 Introduction

Health impact assessments (HIA) is defined as 'a combination of procedures, methods, and tools by which a policy, intervention or service may be judged as to its potential effects on the health of a population, and the distribution of those effects

L. Green (✉)
HIA, WHO Collaborating Centre on 'Investment for Health and Wellbeing', Public Health, Wales, UK
e-mail: Liz.Green@wales.nhs.uk

H. Brand et al. (eds.), *Climate Change and Public Health*,
https://doi.org/10.1007/978-981-95-0895-2_10

within a population' (European Centre for Health Policy 1999). Underpinned by the WHO 'holistic definition of health and wellbeing, it uses the social determinants of health as a lens through which to identify health impacts' (World Health Organization 1948). These are factors such as health behaviours, environmental conditions, socialization factors, and economic factors (Dahlgren and Whitehead 1991) and should be read in conjunction with this paper (Dahlgren and Whitehead 1991). HIA is 'an open, scalable, and transparent evidence-based process and contains several elements for example a literature review, interviews, workshops, health intelligence, and statistical data' (Winkler et al. 2021).

HIA at its core seeks to identify who is going to be affected in the population by a policy or plan and how their wider health and wellbeing will be impacted by it. The impact can be identified as potentially positive or promote opportunities for health and/or negative and identify any unintended consequences for health and wellbeing. If negative, then it looks at how these impacts can be mitigated and how any inequalities in the distribution of impact can be addressed. HIAs can lead to tangible and practical suggested actions, and recommendations to improve and protect health are a key part of the HIA process. HIA has been recommended by several scientific advisory bodies including the WHO (World Health Organization 2021).

10.2 Evolution of HIA Practice

HIA emerged out of the practice of Environmental Impact Assessment (EIA) in the late 1990s. Since this time, HIA has advanced (Birley 2011; Rogerson et al. 2020; Winkler et al. 2020). Recently multiple reviews of HIA practice internationally have been published (Winkler et al. 2020; Cave et al. 2021; Green et al. 2022). These show variability in the use of HIA globally and with some geographic or regional areas carrying out HIAs but others none. At this time, Europe has most HIA activity and there it has progressed so that HIAs are carried out across a wide range of policies, plans, and on project developments (Haigh et al. 2015; Winkler et al. 2020; Cave et al. 2021; Green et al. 2022). Australia and New Zealand and the United States follow (Winkler et al. 2020). The HIAs undertaken can range from standalone HIAs with a primary focus on health and equity (Harris-Roxas and Harris 2007; Haigh et al. 2015; Green et al. 2022) to HIA that focus on the environmental determinants of health and risk to health inclusion in other impact assessments for example, Environmental Impact Assessment (EIA) and Strategic Environmental Assessment (SEA) (Fischer 2008; Cave et al. 2021). HIA in any nation or region is also highly dependent on the context, legislative background, and policy drivers within it (Rogerson et al. 2020; Cave et al. 2021; Green et al. 2022).

The x context (political, economic, and social) and the regulatory or legal frameworks that HIA is carried out within is also important. For example, countries such as Wales have a political and social context (devolution, a progressive government with a social and health focus) and have made HIA a statutory requirement to be carried out in decision and policy making. This can act as a vehicle for enabling HIAP in practice and facilitating integrated policy and plan making in order to reduce health inequalities (Welsh Government 2017). Others, for example, Thailand, have HIA as statutory requirement as part of legislation around community engagement or environmental protection (Chandanachulaka 2012), whilst others states and nations, for example, Quebec and Canada, use HIA as a way to mobilize HIAP and to improve decision-making (Boldo et al. 2011; Government of Canada 2012). In the Indian sub-continent, there are no requirements for HIA and no standalone HIAs have been identified in the peer-reviewed literature. However, some have been identified in the grey literature. At an organizational level, the Asian Development Bank requires HIAs to be carried out when funding is being sought for development projects (Asian Development Bank 1992, 2018). HIA is the main tool to consider health, wellbeing, and equity in non-health sectors in a practical and systematic way. It can promote 'Health in all policies' across sectors and settings, and this includes climate change policy, adaptation, and mitigation plans.

10.3 HIA and Climate Change

Climate change has a wide range of effects on the wider determinants of health, for example, water cycles and the availability of water, temperature and weather extremes, and the integrity and health of ecosystems of which humans are an integral part. It also affects housing and settings such as schools and hospitals when extreme weather events such as flooding or high temperatures occur. Climate change has been recognized to be the biggest threat to global health of the twenty-first century (Costello et al. 2009) making it evident that a healthy planet is a prerequisite for human health. A recent report by leading climate scientists suggests that many of the models assessing the economic risk associated with climate change have been omitting or seriously underestimating many of the most serious consequences for lives and livelihoods (DeFries et al. 2019). COP 26 emphasized the need for climate action (Nations 2021) at a national and global level, and many nations have declared climate emergencies (Welsh Government 2019; World Atlas n.d.). The UN is increasingly concerned about the impact of climate 'boiling' (United Nations 2023). Health and wellbeing are referenced by the Intergovernmental Panel for Climate Change as important and state that mitigations for climate change could provide significant opportunities for population health (IPCC 2021, 2022). Nevertheless, it could also widen inequalities, particularly in the short to medium term. HIA is a beneficial tool to address these, as it can provide evidence to inform mitigation and adaptation decisions, identify those most affected, and take a short to long-term view (Green et al. 2023).

This chapter focuses primarily on a Health Impact Assessment (HIA) that was carried out in Wales which centred around the health, wellbeing, and equity impacts of climate change. The HIA aimed to identify the potential health and wellbeing impacts of climate change, whether they be physical, mental, social, economic, or environmental, using the framework of the wider determinants of health and wellbeing. The chapter will describe the process and the results of the HIA and highlight how HIA is a useful tool for policymakers when thinking about action on climate change. It briefly outlines the quantitative aspects as well as highlights the benefits of qualitative evidence within the process, for example, engagement with stakeholders and interviews with key informants from a range of national organizations. The chapter illustrates how an HIA can help identify the differential effects of climate change on the health and wellbeing of different populations.

10.4 Welsh Context

Wales is a nation within the United Kingdom and is smaller than some other states in countries such as the United States, Australia, or India. Wales has a population of just over 3,000,000 and is predominantly rural with a couple of cities, mainly in the South. It is devolved and, therefore, can implement its legislation in a range of areas, for example, health and social care, the environment, and climate change. It can also make regulations around spatial planning, land use planning, transport, and other areas that are related to climate change or where climate change is relevant.

The Welsh Government has a very clear focus on sustainable development. It enacted the Wellbeing of Future Generations (Wales) Act (2015) and the United Nations SDGs are built into this unique legislation (Welsh Government 2021; Future Generations Commissioner for Wales 2023). It is the first country in the world to legislate in this way for the SDGs. It enforces public bodies in Wales such as the Welsh Government, Public Health Wales, local municipal authorities, the environment agency for Wales—Natural Resources Wales, and other public bodies to consider the long-term impact of their actions on future generations across a range of issues, for example, economy, health, Welsh language and culture, global responsibility and the environment. It also requires bodies to collaborate effectively with people and each other to avoid actions that may cause damage in the short and long term (Welsh Government 2021).

Climate change is an important issue for the Welsh Government to tackle. The country declared a climate emergency in 2019 and appointed a Minister for Climate Change in 2021 (Welsh Government 2019). This was centred around the increased incidence of heat, storms, and flooding that caused widespread damage to Wales from 2019 to 2022 as well as being part of a global responsibility. Alongside, health is also deemed important and there is a clear drive to consider health and wellbeing through the use of Health Impact Assessment, which is statutory as part of the Public Health Wales Act 2017 (Welsh Government 2017).

In 2020, Public Health Wales (PHW) internally commissioned a comprehensive HIA on climate change in Wales. It provided an opportunity to examine and better understand the health and wellbeing impacts of climate change in the future for Wales and in what way they would play out and who would be potentially affected. HIA has been utilized in Wales over several years and across a range of topic areas and led by a range of organizations including PHW. This HIA would be used to help inform PHW and other bodies planning for climate change and support their ambitions under the Wellbeing of the Future Generations (Wales) Act and Public Health (Wales) Acts. These Acts place a duty on public bodies in Wales (Welsh Government 2021) to carry out HIA in specific circumstances and to maximize seven Wellbeing Goals which include the environment, health, equity, global responsibility, and Welsh language and culture. They do this by working together or individually via the '5 ways of working' or the sustainable development principle—long-term thinking, prevention, collaboration, integration, and public involvement (Welsh Government 2017). The HIA was led by the Wales Health Impact Assessment Support Unit (WHIASU) team who are based in the Policy and International Health Directorate of PHW. The team provides advice, guidance, and training in HIA and helps facilitate the process (Edmonds et al. 2019).

10.5 HIA Process

The HIA process followed a 5-stage process, as recommended in much of the HIA literature (Winkler et al. 2021; Green 2023). It gathered and analysed both quantitative and qualitative data and evidence to form a coherent picture of the impact on Wales from climate change. It encompasses screening, scoping, appraisal of the evidence, reporting, review, and reflection including monitoring and evaluation (Table 10.1).

To address the need for governance, a Strategic Advisory Group was established which included some of the key decision-makers in Wales such as the Welsh Government, Natural Resources Wales, Public Health Wales, and health boards and academics from Cardiff University or Aberystwyth University. They critically reviewed the HIA and provided a steer throughout the process.

Table 10.1 Steps followed in the HIA of climate change in Wales

HIA Step	Actions	
Screening	Identified the preliminary health and wellbeing impacts and those affected in the population.	
Scoping	Used a scoping checklist to identify the: Geographical boundaries of the HIA Timeframes/deadlines Terminology to be used to define and characterize impact. Stakeholders Engagement methods Data and evidence sources	
3a. Appraisal – Evidence gathering	Literature review	Carried out literature review and synthesized into summary to identify relevant qualitative and quantitative evidence and statistics
	Collate community health profile and environmental data	Used the scoping and screening checklists as a guide to gather data to identify relevant health intelligence and demographic, economic, environmental, and social data/statistics. This included gathering data in relation to population groups affected and determinants of health identified.
	Stakeholder evidence	Identified key information, knowledge, and evidence of external stakeholders identified as part of the scoping process by: Undertaking interviews and transcribing for analysis. Carried out two participatory stakeholder workshops. Developed a summary of results to input into the final report.
3b. Appraisal of evidence	Triangulated evidence to assess and characterize the positive and negative impacts and form a picture of the scale, scope, and duration of these. Made suggestions for action and main findings in dialogue with stakeholders.	
4. Reporting and recommendations	Compiled report and suggested actions. Carried out quality assurance with key stakeholders. Reviewed and approved in PHW. Published HIA.	
5. Review, reflection, and monitoring	Reviewed and evaluated the process of carrying out the HIA. Developed monitoring indicators/steps.	

(Adapted from Edmonds and Green 2023)

10.6 Conducting the HIA

The HIA combined a desk-based literature review with the collection of primary data through qualitative methods to synthesize all available evidence on the health effects of climate change in Wales.

A research protocol was developed for a rapid review of both grey and academic peer-reviewed evidence and had explicit inclusion and exclusion criteria. There were no restrictions on date or countries, and the key search terms were 'climate change' OR 'global warming' OR 'extreme weather' OR 'flood' OR 'systematic review' OR 'meta-analysis' plus a combination of them all. Additionally, a wide range of wider determinants terms, for example, displacement, economy, employment, and groups such as agricultural workers, farmers, and fishers. ProQuest, Medline, and Web of Science were searched. All results were screened and reviewed by four independent researchers. If deemed eligible for inclusion, appropriate information was pulled into an extraction table for use in the evidence appraisal process.

A community health profile was developed using quantitative health intelligence data which was obtained from robust evidence sources in Wales. This included the National Survey for Wales (Welsh Government 2020), the Office of National Statistics demographic data (Office for National Statistics 2023), and the Welsh Index of Multiple Deprivation (Stats Wales 2020). Quantitative data was compiled from the Climate Change Risk Assessment (CCRA) for Wales (Kovats and Brisley 2021) and Natural Resources Wales flood maps, as well as other environmental and climatic data (Holbrook et al. 2020; Kovats and Brisley 2021; IPCC 2022; Natural Resources Wales 2022)).

Participation and stakeholder involvement is strongly recommended as part of the HIA process (Mahoney et al. 2007; Green 2015; Haigh et al. 2015; Winkler et al. 2021) to ensure that the process is inclusive and qualitative evidence is also gathered. In total, 25 stakeholders were invited to participate in semi-structured qualitative interviews, with a total of 19 being completed. In addition, two cross-sector, multidisciplinary participatory stakeholder workshops were also held. Overall, 50 stakeholders were invited to participate in the workshops, which resulted in 33 individuals agreeing to attend. The purpose of the workshops was to gather qualitative evidence and knowledge from key stakeholders about the potential impacts of climate change. Participating stakeholders were identified purposively based on the findings from the screening stage and included representatives from a wide range of public bodies, academia, and the third sector (civil society), for example, Welsh Government (Welsh Government n.d.-a), Natural Resources Wales (Natural Resources Wales 2023) (the environment agency), Local Health Boards (Welsh Government n.d.-b) (who provide healthcare services), Cardiff University, Renew Wales (EGIN n.d.), and local municipal government agencies (WLGA n.d.). The stakeholders were cross-sectoral, for example, health and social care, housing, planning, transport, schools, and community representatives. The workshops were independently facilitated by members of the WHIASU team. No formal ethical approval was required to undertake this assessment, and all participants agreed to findings being reported anonymously and consent was taken before participation. Workshop participants also completed a feedback and evaluation form.

The two workshops focused on different levels of government and the people for each level were identified. The first workshop provided strategic and policy perspectives along with organizational community insights. The second workshop focused on the impacts at a local level on the actual population, for example,

Fig. 10.1 Map of Wales and Fairbourne. (Source: Jenney Creative, 2023 and Mark Drane, 2021)

analysing the impact of flood events on people and the ramifications on their settings or within their organizational settings such as schools, or at the local community level. Representatives of organizations included housing, transport, sustainability, public health at a local level, and construction. Community and voluntary organizations in schools were invited as they are at the forefront of helping and supporting communities in distress.

The interviewees selected were specifically experts in their field belonging to the fields of climate change, travel, and active transport at the sociological level or the environmental risk factor level. Further, two case studies were conducted—one of which was a short Rapid HIA on the community of Fairbourne in North Wales, which is a local authority, area of Gwynedd (see Fig. 10.1). It drew attention to the causes, the current scenario, and proposals for mitigation/adaptation to deal with predictions for extreme weather events in the next 30 to 40 years.

10.7 Health Impacts of Climate Change in Wales

The HIA findings indicate significant potential impacts across the wider determinants of health and mental wellbeing (Table 10.2). Across the determinants of health, one of the major findings is that climate change will potentially negatively affect people's mental health and wellbeing. Stress, anxiety, and uncertainty about

Table 10.2 Determinants of health affected by climate change in Wales. (Adapted from Edmonds and Green 2023)

Determinants of Health (D) impacted by climate change in Wales (A = additional areas that require more evidence)	Positive impacts/ opportunities	Negative impacts
Food security and nutrition	✓	✓
Behaviours affecting health		
Physical and outdoor leisure activity	✓	✓
Alcohol and substance misuse		
Social and community factors		
Community resilience and cohesion	✓	✓
Population displacement, mobility, and migration	✓	✓
Violence		
Family and intergenerational relationships		
Mental health and wellbeing	✓	✓
Living and environmental conditions affecting health		
Housing	✓	✓
Air quality	✓	✓
Flooding		✓
Higher temperatures and extreme heat	✓	✓
Water supply and quality		✓
Natural environment and biodiversity	✓	✓
Landslides and coal tips		✓
Wildfires		✓
Vector borne disease		✓
Economic conditions affecting health		
Working conditions	✓	✓
Economic development and skills	✓	✓
Access and quality of services		
Health and social care: Access and delivery	✓	✓
Education and skills	✓	✓
Macro-economic, environmental and sustainability factors		
Transport	✓	✓
Infrastructure	✓	✓

the future along with lack of control, participation, and resilience are key factors affecting mental health and wellbeing (Amelung et al. 2019; Kovats and Brisley 2021). However, there were also examples of people and communities coming together and supporting each other during extreme weather events, which can be considered a positive impact, contributing to community resilience and cohesion.

Other identified impacts included food and nutritional security. Wales imports and exports a lot of its products, putting it at risk of negative impacts from climate events even elsewhere in the world. A tsunami on one side of the world, for example, could disrupt transport, travel, and supply lines. Wales also has a large lamb farming community that produces wool, and climate change could potentially affect the livelihoods of these farmers and could lead to increased poverty.

Furthermore, the HIA highlighted that climate change and the environment could provide a chasm between generations due to a different frame of thinking around this along with other issues. This has been explored more in another piece of work looking at the cumulative impact of Brexit (the UK withdrawal from the EU), the pandemic, and climate change. For example, for Brexit, younger people tended to vote to remain in the European Union, whereas many older people voted to withdraw and leave the European Union. It can be seen clearly in voting patterns there, which can be witnessed around discussions on climate change. The health and wellbeing impacts of Brexit, COVID, and climate change are happening together and are a triple challenge (Green et al. 2021).

The HIA also highlighted that there will be moderate to major negative impacts in the short and long term around the environment, economy, and health behaviours. Although there could also be minimal to moderate opportunities for health and wellbeing, there could be significant opportunities in the long term, for example, by reducing energy use, focusing on circular economy approaches, and utilizing the Wellbeing of the Future Generations (Wales) Act which promotes long term thinking and public bodies working together to tackle significant issues such as climate change or health and the environment (Welsh Government 2021).

10.8 Vulnerability to Climate Impacts in Wales

In terms of vulnerable groups, there is a wide range of potential population groups, occupations, and settings like the farmers, children and young people, schools, care homes, and prisons affected both in the short and long term (Table 10.3). For example, farmers and agricultural workers are going to be severely affected both at an agricultural and economic level. Another group that will be affected are those who need healthcare and their caregivers as services must continue despite any extreme weather events. The impacts identified are both confirmed (current) and potential.

Table 10.3 Population groups affected by climate change in Wales. (Adapted from Edmonds and Green 2023)

Population Groups who are more vulnerable and disproportionately impacted by the health and wellbeing impacts of climate change in Wales	Positive impacts/ opportunities	Negative impacts
Age-related groups		
Babies, children, and young people	✓	✓
Older adults	✓	✓
Sex/gender-related groups		
Women		✓
Pregnant women		✓
Men	✓	✓
Groups at higher risk of discrimination or another social disadvantage		
Displaced people and refugee and asylum seekers (see D3.2)		✓
People with long-term health conditions and/or disabilities		✓
People who are homeless		✓
Ethnic minority groups		✓
People who are new to an area		✓
Income related groups		
People living on a low income	✓	✓
Occupational groups (see D6.1)		✓
Geographical groups		
Coastal communities	✓	✓
Flood risk areas (see D5.3)		✓
Former and current industrial areas	✓	✓
Urban areas	✓	✓
Rural areas	✓	✓
Areas of multiple disadvantage		✓

10.9 Implications of HIA for Policy and Decision-Making

The work has identified a wide range of short- and long-term impacts and that climate change is complex and is an important determinant of health and wellbeing. Some of these are potential opportunities, but many are negative impacts or unintended negative impacts which positive action in one sector, for example, the environment could lead to others, the economy or health.

The work has highlighted that climate change is here now and affecting health and wellbeing and communities and is not something in the future (IPCC 2021; NASA 2023). However, that depends on the policy direction as changes and adaption to some of the policies can transform the country into a low-carbon economy. There is a decarbonization plan for Wales for the NHS for public bodies (Powlesland and Joyce 2021; WG 2022). But that shift could have some short-term negative impacts, the management of which will depend on existing policies, plans, and projects. Through ongoing policy dialogue, HIA can help inform people and their thinking.

It has also flagged that there are many synergies across sectors to promote sustainable development and there are co-benefits of working together. The Wellbeing of Future Generations (Wales) Act 2015, with a focus on seven Wellbeing Goals along with the enabling environment and political will in Wales, are important drivers to promote intersectoral action. For example, health, social care, and environmental agencies can work together when planning services and adapting sites. Other clear goals that require collaboration and coordination for practical implementation on the ground include enhancing adaptive capacity via long-term investment in preventive action, community engagement via communications and education, investing in co-benefits for health, increasing social participation, and carrying out further research such as on demand for health services arising from climate change.

The qualitative evidence and the way it was captured using workshops also provided a forum for discussion, and a wide range of perspectives could be brought together to discuss climate change and health systematically. It has specifically captured which sectors and settings are likely to be most impacted, which most of the existing literature does not. The use of HIA has not only added to the evidence base but also highlighted that it is a beneficial advocacy tool to support health and wellbeing outcomes and health policies and demonstrate that health and wellbeing are integral to climate change and vice versa. The evidence included in the HIA has promoted evidence-based policy planning, and the nature of the evidence has led to capturing a wide range of local and national issues that data and statistical evidence may have missed. For example, qualitative evidence and contextual knowledge are important for identifying the impact on not only individual health and wellbeing but also the organizations, communities, and settings and what climate change will mean for them, their work, leisure, and daily lives. Evidence is crucial to support policies, plans, and projects going forward and the HIA has proposed integrating health and wellbeing into adaptation plans and policymaking for future transitions to low-carbon, net-zero economies.

Part of this work was carried out before the pandemic and finished in 2023. It has been used to inform Public Health Wales since 2022 despite not being completed during the development of its Long-Term Strategy (Public Health Wales 2023), and climate change is now a key strategic priority for the organization. It has also informed and influenced the Welsh Government's climate change work, been promoted in the Chief Medical Officers report in 2022 that included the set of infographics published in time for the UN Climate Change Conference in Glasgow (COP26) reiterated the emergency of climate change (Edmonds and Green 2023) and the impact on health (Nations 2021) which COP 27 reflected with discussions around the health and wellbeing impacts on equity and population groups.

However, there are limitations to the work and challenges faced. It was a hugely complex piece of work that stopped during the pandemic and then restarted in 2022. The evidence base is always evolving and needs to be updated. It also only looks at climate change now and does not include mitigation or adaptation as part of its remit as there are so many complex interventions at play. There are a huge number of potential impacts across the determinants of health and population groups and so it needed to be articulated and communicated well—reflecting the uncertainties, complexities, and challenges in an accurate and meaningful way.

Despite these limitations, this work has benefited policy and decision-makers by providing an up-to-date resource that takes a holistic view of the issues and how these may be addressed. It promotes further use of HIA as a beneficial tool for engagement and discussion with those who are directly making the plans and directly coming up with the policies in Welsh Government or Public Health Wales or the environment agencies and health boards.

10.10 Conclusion

HIA is a practical and useful tool to use in the development of climate change plans and policies in all kinds of settings. However, the practice of HIA is highly dependent on a range of political, environmental, and regulatory factors. Similarly, climate change as a critical policy is complex and not all nations have committed to net zero or recognized the impact of global warming. This HIA has demonstrated how it can add value, facilitate conversations, and raise awareness of the importance of climate change in promoting or harming health and wellbeing. When used systematically, HIA can serve as an important policy tool to support evidence-based and participatory decision-making on complex, intersectoral issues related to climate change mitigation and adaptation.

Acknowledgments Thanks to Kath Ashton and Nerys Edmonds and all who participated in and contributed to collation and writing of the HIA of climate change in Wales.

References

Amelung D, et al. Human health as a motivator for climate change mitigation: results from four European high-income countries. Glob Environ Chang. 2019;57:101918. Available at:. https://doi.org/10.1016/j.gloenvcha.2019.05.002.

Asian Development Bank (1992) Guidelines for the health impact assessment of development projects. Available at: https://www.adb.org/sites/default/files/institutional-document/212016/guidelines-hia-devprojects.pdf. Accessed 15 Feb 2023.

Asian Development Bank Health impact assessment: a good practice sourcebook. 2018. Available at: https://www.adb.org/sites/default/files/institutional-document/452951/health-impact-assessmentsourcebook.pdf. Accessed 15 Feb 2023.

Birley M. Health impact assessment: principles and practice. London: Earthscan; 2011.

Boldo E, et al. The institutionalization of health impact assessment in Quebec: a feasible model in Spain? Gac Sanit. 2011;25(6):535–41. Available at:. https://doi.org/10.1016/j.gaceta.2011.05.007.

Cave B, et al. Lessons from an international initiative to set and share good practice on human health in environmental impact assessment. Int J Environ Res Public Health. 2021;18(4):1392. https://doi.org/10.3390/ijerph18041392.

Chandanachulaka S. Health impact assessment in Thailand. In: Health impact assessment: past achievement, current understanding, and future progress. Oxford: Oxford University Press; 2012.

Costello Costello A, Abbas M, Allen A, Ball S, Bell S, Bellamy R, et al. Managing the health effects of climate change. The Lancet. 2009;373(9676):1693–733. https://doi.org/10.1016/s0140-6736(09)60935-1.

Dahlgren G, Whitehead M. Policies and strategies to promote social equity in health. Background document to WHO – strategy paper for Europe. Arbetsrapport. 1991; [Preprint]. Available at: https://ideas.repec.org//p/hhs/ifswps/2007_014.html. Accessed 16 Feb 2023.

DeFries R, Edenhofer O, Halliday A, Heal G, Lenton T, Puma M, et al. The missing economic risks in assessments of climate change impacts. In: Grantham Research Institute on Climate Change and the Environment; 2019. https://doi.org/10.7916/d8-6f8h-md45.

Edmonds N, Green L. Climate change in Wales: health impact assessment. Summary Report; 2023. Available at: https://phw.nhs.wales/news/health-impact-assessment-highlights-urgent-need-to-protect-health-and-wellbeing-as-the-climate-changes/climate-change-in-wales-health-impact-assessment/

Edmonds N, Parry-Williams L Green L. WHIASU training and capacity building framework for HIA in Wales. 2019. Available at: https://phwwhocc.co.uk/wp-content/uploads/2021/02/WHIASU_Training_and_Capacity_Building_Framework_for_HIA_in_Wales_2019-_2024_FINAL_TECHNICAL_DOCUMENT.pdf. Accessed 15 Feb 2023.

EGIN. The renew Wales programme. EGIN; n.d.. Retrieved September 17, 2023, from https://egin.org.uk/learn-more/renew-wales/

European Centre for Health Policy. Health impact assessment: main concepts and suggested approach (Gothenburg Consensus Paper). 1999. Available at: https://web.archive.org/web/20061007033923/http:/www.who.dk/document/PAE/Gothenburgpaper.pdf. Accessed 16 Feb 2023.

Fischer TB. Theory and practice of strategic environmental assessment. London: Earthscan; 2008.

Future Generations Commissioner for Wales. Well-being of future generations (Wales) Act 2015– the future generations commissioner for wales, future generations commissioner for Wales. 2023. Available at: https://www.futuregenerations.wales/about-us/future-generations-act/. Accessed 22 Mar 2023.

Government of Canada, C.I. of H.R. Guide to knowledge translation planning at CIHR: integrated and end-of-grant approaches. CIHR; 2012. Available at http://www.cihr-irsc.gc.ca/e/45321.html. Accessed 14 Aug 2018.

Green L. The importance of health impact assessments. In: The human and environmental impact of fracking; how fracturing shale for gas effects us and our world. Santa Barbara: Praeger; 2015. p. 49–60.

Green L. Health impact assessment (HIA) as a tool to mobilise 'health in all policies'. Maastricht University; 2023.

Green L, et al. Rising to the triple challenge of Brexit, COVID-19 and climate change for health, well-being and equity in Wales contents. Public Health Wales NHS Trust; 2021. p. 1–56.

Green L, et al. Predicted and observed impacts of COVID-19 lockdowns: two health impact assessments in Scotland and Wales. Health Promot Int. 2022;37(6):daac134. https://doi.org/10.1093/heapro/daac134.

Green L, et al. Brexit, COVID-19 and climate change: mapping the impact of the 'triple challenge' on health and well-being in Wales. Impact Assess Proj Apprais. 2023;41(2):126–38.

Haigh F, et al. What makes health impact assessments successful? Factors contributing to effectiveness in Australia and New Zealand. BMC Public Health. 2015;15:1009. https://doi.org/10.1186/s12889-015-2319-8.

Harris-Roxas BF, Harris PJ. Learning by doing: the value of case studies of health impact assessment. N S W Public Health Bull. 2007;18(9–10):161–3. https://doi.org/10.1071/nb07110.

Holbrook NJ, et al. Keeping pace with marine heatwaves. Nat Rev Earth Environ. 2020;1(9):482–93. https://doi.org/10.1038/s43017-020-0068-4.

IPCC. Climate change widespread, rapid, and intensifying–IPCC — IPCC. IPCC; 2021. Available at https://www.ipcc.ch/2021/08/09/ar6-wg1-20210809-pr/. Accessed: 21 Mar 2023.

IPCC, editor. 'Summary for policymakers', in global warming of 1.5°C: IPCC special report on impacts of global warming of 1.5°C above pre-industrial levels in context of strengthening response to climate change, sustainable development, and efforts to eradicate poverty. Cambridge: Cambridge University Press; 2022. p. 1–24. https://doi.org/10.1017/9781009157940.001.

Kovats S, Brisley R. The third UK climate change risk assessment technical report. 2021. Available at: https://www.ukclimaterisk.org/independent-assessment-ccra3/technical-report/. Accessed 27 July 2023.

Mahoney ME, Potter J-LL, Marsh RS. Community participation in HIA: discords in teleology and terminology. Crit Public Health. 2007;17(3):229–41. https://doi.org/10.1080/09581590601080953.

NASA. Evidence | facts—climate change: vital signs of the planet. NASA; 2023. Available at https://climate.nasa.gov/evidence/. Accessed 21 Mar 2023.

Natural Resources Wales. Natural resources Wales/Wales facing a rising climate tide. 2022. Available at: https://naturalresources.wales/about-us/news/news/wales-facing-a-rising-climate-tide/?lang=en. Accessed 8 Mar 2023.

Natural Resources Wales. Natural resources Wales. Natural Resources Wales; 2023. https://naturalresourceswales.gov.uk/?lang=en

Office for National Statistics. General census data. 2023. Available at https://www.ons.gov.uk/.

Powlesland D, Joyce C. NHS Wales Decarbonisation strategic delivery plan. Welsh Government [Preprint]. 2021.

Public Health Wales. Long Term Strategy 2023–2035. 2023. Available at: https://phw.nhs.wales/about-us/working-together-for-a-healthier-wales/phw-strategy-plain-eng-0523-pdf/.

Rogerson B, et al. Recent advances in health impact assessment and health in all policies implementation: lessons from an international convening in Barcelona. Int J Environ Res Public Health. 2020;17(21):7714. https://doi.org/10.3390/ijerph17217714.

Stats Wales. Welsh Index of Multiple Deprivation. 2020. Available at: https://statswales.gov.wales/Catalogue/Community-Safety-and-Social-Inclusion/Welsh-Index-of-Multiple-Deprivation. Accessed 13 Apr 2023.

United Nations. COP26: together for our planet. United Nations; 2021.

United Nations. Hottest July ever signals 'era of global boiling has arrived' says UN chief. 2023. Available at: https://news.un.org/en/story/2023/07/1139162.

Welsh Government. Public Health (Wales) Act 2017. 2017. Available at: https://www.legislation.gov.uk/anaw/2017/2/contents/enacted. Accessed 16 Feb 2023.

Welsh Government. Welsh Government makes climate emergency declaration, GOV. WALES. 2019. Available at: https://www.gov.wales/welsh-government-makes-climate-emergency-declaration. Accessed 8 Mar 2023.

Welsh Government. National Survey for Wales: monthly survey. 2020. Available at: https://gov.wales/sites/default/files/statistics-and-research/2020-06/national-survey-for-wales-monthly-survey-may-2020_0.pdf. Accessed 20 Apr 2023.

Welsh Government. Well-being of future generations (Wales) act 2015 essentials guide. Welsh Government. 2021; [Preprint]

Welsh Government. Home | GOV.Wales. Welsh Government Services and Information; n.d.-a. Retrieved September 17, 2023, from https://www.gov.wales/

Welsh Government. NHS Wales health boards and trusts | GOV.Wales. Welsh Government; n.d.-b. Retrieved September 17, 2023, from https://www.gov.wales/nhs-wales-health-boards-and-trusts

WG. Net Zero Strategic Plan. Welsh Government; 2022. p. 1–49.

Winkler MS, et al. Current global health impact assessment practice. Int J Environ Res Public Health. 2020;17(9):2988. https://doi.org/10.3390/ijerph17092988.

Winkler, M. et al. Health impact assessment international best practice principles. Special publication series no. 5. 2021. Available at: https://www.researchgate.net/publication/352573139_Health_impact_assessment_international_best_practice_principles_International_Association_for_Impact_Assessment. Accessed 15 Feb 2023.

WLGA. Local government in Wales – WLGA. Welsh Government Association; n.d.. Retrieved September 17, 2023, from https://www.wlga.wales/local-government-in-wales

World Atlas. Which countries have declared climate emergencies? n.d.. Available at: https://www.worldatlas.com/articles/which-countries-have-declared-climate-emergencies.html.

World Health Organization. Constitution of the World Health Organization. 1948. Available at: https://www.who.int/governance/eb/who_constitution_en.pdf. Accessed 15 Feb 2023.

World Health Organization. Health impact assessment. 2021. Available at: https://www.who.int/health-topics/health-impact-assessment. Accessed 16 Feb 2023.

Public Health and Climate Change: The Israeli Perspective

11

Nadav Davidovitch

Abstract

Israel is experiencing the effects of climate change, with temperatures projected to rise by about four degrees by the end of the century. The country is expected to suffer more than northern countries due to higher temperatures, increased humidity, rising sea levels, and extreme weather fluctuations, which will cause an increase in the risk of natural disasters, famine, floods, water contamination, forced migration, pandemics, and increased border tensions. However, Israel's unique position as a country between three continents, with an advanced healthcare system under the National Health Insurance Law and a strong scientific sector, serves as a foundation for developing innovative solutions. This chapter delves into heat-related emergencies in the country, strategies for coping with the current situation and potential future threats (including climate migration), lessons learned from the COVID-19 pandemic, and the role of institutions and other actors in global and regional cooperation. It argues that there is a dire need to consider the influence of climate change on health from a broader eco-health perspective.

Keywords

Climate migration · Antimicrobial resistance · Israel · COVID-19 · Regional resilience · Cross-border collaboration

N. Davidovitch (✉)
School of Public Health, Ben-Gurion University, Negev, Israel
e-mail: nadavd@bgu.ac.il

H. Brand et al. (eds.), *Climate Change and Public Health*,
https://doi.org/10.1007/978-981-95-0895-2_11

According to the World Meteorological Organization (WMO), "The State of the Climate" report draws attention to the increase in GHG and accumulated heat that has led to the hottest weather recorded in the past 8 years (World Meteorological Organization 2022). Israel is suffering from climate change with a future rise by the end of the century of about four degrees (Environment and Health Fund 2017). Since the 1950s, the annual mean temperature in the country has risen by 1.7 °C, with the summer leading the way at a rate of 0.27 °C per decade (over the period 1950–2020) (Lev and Yosaf 2023). It is going to suffer more than northern countries with higher temperatures, higher humidity, rising sea levels, and more extreme weather fluctuations that will cause an increase in the risk of natural disasters, famine, floods, water contamination, forced migration, pandemics, and increased border tensions. On the other hand, the unique situation of Israel, as a country in between three continents, with an advanced healthcare system based on the National Health Insurance Law, and with a strong scientific sector—all serve as a foundation for developing innovative solutions. Israel is a high-income country with both ongoing economic growth but also growing inequalities. Israel's climate differs across the country between Mediterranean, arid, and semi-arid climates. According to the WHO Report on Health and Climate Change—Country Profile (2022), Israel is prone to a variety of health risks including health impacts of extreme weather events, respiratory illnesses, waterborne diseases, Zoonoses and VBDs, malnutrition, foodborne diseases as well as noncommunicable diseases, mental, and psycho-social health.

Premature death from exposure to extreme heat has risen during the last decade. The former Minister of Environmental Protection Tamar Zandberg stated, "The climate crisis is a real threat to health. The study shows the impact of the heat wave and the climate crisis on excess mortality, especially on weaker populations" (Ministry of Environmental Protection 2022). According to the WHO Report on Health and Climate Change mortality risk between 1998 to 2014, after 6 days the mortality risk in three age groups (65+, 65–74, and 75+ years) was 1.07, 1.15, and 1.05, respectively. Looking at heat waves in Israel from 2012 to 2020 on average 45 people die per year due to heat waves, with those over the age of 70 being the most affected by the sudden temperature changes (Ministry of Environmental Protection 2022).

A rise in heat-related emergency room visits has been observed. It includes the number of hospitalizations, the length of stay, and significant increases in healthcare costs. Cardiovascular as well as cerebra-vascular diseases, respiratory and other diseases, all are causes of morbidity and mortality including higher healthcare utilization demands including hospitalization. Furthermore, studies from around the world and from Israel show that heat waves have a disproportionate impact on hospitalizations and mortality in the elderly, women, and low-income populations (Ministry of Environmental Protection 2022). A study from Soroka Medical Centre located in the Southern area of Israel called Negev experiences dusty periods, leading to air pollution including high rates of PM 2.5, and there was an increased risk of hospitalization due to respiratory illnesses such as Chronic Obstructive Pulmonary Disease (COPD) and asthma exacerbation. Another Israeli study found that high ambient temperatures were associated with increased stroke risk (Vered et al. 2020).

All these studies reflect similar findings from other countries. An important question to be asked is the way of coping with the current situation and potential threats in the future. An important lesson should be drawn from the COVID pandemic. With the onset of the COVID pandemic, joint multi-sectoral efforts to mitigate the pandemic were observed. The COVID pandemic mobilized many countries; severe measures were taken to curb the spread including lockdowns and cancelling flights. Why is such mobilization not taking place in the context of climate change? Even more remarkable to note is that despite challenges and far from perfect response, data from a variety of sources became available. It is interesting to note that this call is not new. Prof Manfred Green, former director of the Israeli Centre for Disease Control already in 2013, raised the issue of the lack of data and the need for equipping existing systems to monitor deaths that would help in analysing climate change and its aspects and propose an effective intervention. Only recently, after more than a decade and probably after the COVID experience, the Israeli Ministry of Health together with other ministries, is now working on a comprehensive approach to promote data integration for action. Learning from the COVID response to boost climate change efforts should be one of the priorities.

The Lancet Countdown report similarly draws attention to the progress observed in countries regarding the health sector; however, application services to policymaking are low. Only a few Member States within the EU have climate services as guiding health sector policy decisions and investment plans. In this section of the health sector in climate change, the response is slow as observed in national adaptation and national health plans as many of them are yet to be adopted, and integration and healthcare system adaptations are still poor. This slow response can be related to the paradox of prevention. While clinical care is responsible for at most about 20 percent of health outcomes, the rest is related to human behaviours, social and economic factors, and the physical environment. However, this is the situation most of the health system budget is invested downstream in medical services. Similarly, in the case of climate change, there is a gap between the alarming data and the investment for priorities within the system. Moreover, in many ways, the current prioritization is harming our public health even contrary to clinical thinking, as climate change effects have a tremendous impact on our health as well as on our wellbeing. We must move from individual thinking related mainly to biological factors and even lifestyle to broader issues related to health in all policies to rethink the local economy including issues related to food and nutrition. Both the built environment and the natural environment up to the global ecosystem are crucial here. The COVID-19 crisis revealed the challenges regarding policies and reducing inequalities. These urgent lessons must be adapted to climate change response. There are many opportunities for action: We need to strengthen the implementation of Israel's National Health and Climate Change Strategy which is currently in a very initial state. Israel must conduct a climate change and health vulnerability and adaptation assessment, assess health co-benefits of national climate mitigation benefits including their measurements, and build climate-resilient and environmentally sustainable healthcare facilities. Many of these are promoted now by the Ministry of Health, but there is a need for more sustainable support including from the Prime Minister's

Office. In October 2021, Israeli President Herzog announced the establishment of the Israeli Climate Forum. One of the subgroups is dealing with health and welfare issues, with a strong emphasis on vulnerable groups.

There is a dire need to consider the influence of climate change on health from a broader eco-health perspective. The interaction between anthropogenic and natural forces is dynamic and critical for our purposes. There is a decline in natural resources versus population growth, and climate change is one of the major forces affecting migration, urban planning, and economic development. This crisis can also serve as a wonderful opportunity to promote public health. A shared perspective on the study of human population and planet Earth must be understood as dynamic and focuses on the interaction between the biotic and abiotic elements, bringing global and multidisciplinary perspectives and their influence on the local context. Moreover, it must include the idea of the need for an integrated assessment that moves from science to policy. This perspective requires the need to move from rhetoric into practice and how to structure the system in a different way, that is, the need to move from evidence-based medicine to evidence-informed health policy, and moving from a more clinical individualistic-oriented approach to an eco-health approach.

The Ben Gurion University has opened a new Centre for Knowledge Transfer and Exchange (KTE) to have a dynamic and iterative process of synthesis, dissemination, and ethically sound application of knowledge by all stakeholders. In addition, climate change was taken as one of the case studies. It is not about producing new knowledge, but also about informing policy, systematic reviews using also local evidence, different types of evidence, great literature, and application of knowledge across different disciplines, and crossing the silos that are many times preventing researchers to integrate different kinds of knowledge.

One example that is also relevant to climate change is the new emerging diseases such as antimicrobial resistance. When complex problems such as climate change or climate change, for example, related to infectious diseases and antimicrobial resistance arise, there is a need to be wary of simple solutions. For example, vaccination cannot be the only solution for COVID. There is a need for multifaceted approaches as mentioned in the previous chapters. In the case of migration, there is an ongoing public health challenge all around the world, including in Israel. There is a growing discourse around environmental refugees because of climate change. Their environmental vulnerability has created an upward trend in migration (Lubanov 2018). A health impact assessment study was conducted in Gaza, to study environmental refugees and understand migration as a socio-ecological determinant of health and the influence of climate change and different political and social drivers to understand how the healthcare system can deal with migration. In addition, this has been a permanent issue throughout human history. There are different driving forces and pull factors and mobility is increasing and must be taken into consideration. Climate change also creates internal migration as people look for better places for better job opportunities as seen in Africa. Moreover, people are moving from one country to the other as seen in Israel, for example, migrant workers and asylum seekers from Eritrea and Sudan.

Climate change migration creates many challenges such as the exclusion of migrants, different barriers, and the price of seeking healthcare (Schwerdtle et al. 2017). It is a life course perspective from infancy, pregnancy, and exposure to pollution and exposure to environmental effects in different epigenetic factors that are working up to the process of migration (Abubakar et al. 2018; Lubanov 2018; McMichael 2020; Schütte et al. 2018). For example, moving to the new country affects the second generation and the off springs (Schütte et al. 2018). Different environmental stressors, including heat waves, water shortage, floods, desertification, and sea level rise, can bring different health outcomes that were mentioned, including displacements because of food security, energy, poverty, rising different vectors, or leading to poverty and health. This is crucial in places like Sub-Saharan Africa and many of them migrate to countries such as Israel as there are more affluent countries. Internal migration in India creates lots of issues as it is a vast country (Bharadwaj et al. 2021; Khan and Arokkiaraj 2021). There is a requirement for a long-term strategy to strengthen the health system in, for example, poor countries like in Africa but also in receiving countries like Israel. There is a need to focus on policy-oriented research, how to develop hospitals, train professionals about climate change, reduce mortality, coordinate care between hospitals and community, and have a comprehensive healthcare system within the country and a country with culturally appropriate services as immigration is bound to increase.

Academic institutions have a very important role in terms of promoting both the understanding of pathways related to climate change and health as well as promoting technological innovations. Yet, the social and political contexts are no less important. In a collaboration of different scholars from Israel and the region, we tried to focus on the lessons learned from COVID-19, moving from a national-based response to a regional and international response (Negev et al. 2021). Although diseases have no borders in climate change, in actual life political processes construct borders. Thus, political will is crucial when tackling global health threats that transcend political borders. The COVID-19 outbreak in Israel, Palestine, and Jordan affected regional demography and politics. The lessons from the crisis demanded better preparedness for climate change impacts and the requirement for stronger economies to take a leading role in developing regional resilience and regional cooperation is crucial (Dahdal et al. 2021).

Promoting regional cooperation through cross-border collaboration is difficult in the current political situation; however, a bottom-up approach can be employed. Many diseases are climate sensitive, and climate change is affecting human health through different pathways. This can be also an opportunity to work across the sector. This is also the whole idea behind public health. In addition, it is a challenge, but it can also help in building trust and solidarity.

References

Abubakar I, Aldridge R, Devakumar D, Orcutt M, Burns R, Barreto M. The UCL–Lancet Commission on migration and health: the health of a world on the move. Lancet. 2018;392:2606–54.

Bharadwaj R, Hazra SH, Das S, Kaur D. Connecting the dots: climate change, migration and social protection – India | ReliefWeb. IIED; 2021. https://reliefweb.int/report/india/connecting-dots-climate-change-migration-and-social-protection

Dahdal Y, Davidovitch N, Gilmont M, Lezaun J, Negev M, Sandler D, Shaheen M. Lessons of the Israeli-Palestinian conflict for public health: the case of the COVID-19 vaccination gap. Int J Environ Res Public Health. 2021;18(21):11292. https://doi.org/10.3390/IJERPH182111292.

Environment and Health Fund. Climate change. Environment and Health Fund; 2017. https://www.ehf.org.il/en/climate_change

Khan A, Arokkiaraj H. Challenges of reverse migration in India: a comparative study of internal and international migrant workers in the post-COVID economy. Comp Migr Stud. 2021;9(1):1–19. https://doi.org/10.1186/S40878-021-00260-2/TABLES/1.

Lev L, Yosaf Y. State of the climate report 2022. Israel Meteorological Service; 2023. https://ims.gov.il/en/node/2113

Lubanov C. Migration and climate refugees in the Middle East and Mediterranean region. The Association of Environmental Justice in Israel (AEJI); 2018. http://www.aeji.org.il/en/content/migration-and-climate-refugees-middle-east-and-mediterranean-region

McMichael C. Human mobility, climate change, and health: unpacking the connections. The Lancet Planetary Health. 2020;4(6):e217–8. https://doi.org/10.1016/S2542-5196(20)30125-X.

Ministry of Environmental Protection. The Ministry of Environmental Protection studied for the first time the impact of heat waves and the climate on mortality in Israel: every heat wave in Israel leads on average to excess mortality of about 45 people; particularly over the age of 70 | Ministry of Environmental Protection. Government of Israel. 2022. https://www.gov.il/en/departments/news/moep_research_on_heat_waves_climate_crisis_and_mortality_in_israel

Negev M, Dahdal Y, Khreis H, Hochman A, Shaheen M, Jaghbir MTA, Alpert P, Levine H, Davidovitch N. Regional lessons from the COVID-19 outbreak in the Middle East: from infectious diseases to climate change adaptation. Sci Total Environ. 2021;768:144434. https://doi.org/10.1016/j.scitotenv.2020.144434.

Schütte S, Gemenne F, Zaman M, Flahault A, Depoux A. Connecting planetary health, climate change, and migration. Lancet Planet Health. 2018;2:e58–9.

Schwerdtle P, Bowen K, McMichael C. The health impacts of climate-related migration. BMC Med. 2017;16(1):1–7. https://doi.org/10.1186/S12916-017-0981-7/FIGURES/1.

Vered S, Paz S, Negev M, Tanne D, Zucker I, Weinstein G. High ambient temperature in summer and risk of stroke or transient ischemic attack: a national study in Israel. Environ Res. 2020;187:109678. https://doi.org/10.1016/J.ENVRES.2020.109678.

World Meteorological Organization. Provisional state of the global climate in 2022. World Meteorological Organization; 2022. https://public.wmo.int/en/our-mandate/climate/wmo-statement-state-of-global-climate

Climate Diplomacy as a Tool for Climate Resilience in Small Island Developing States

12

Kiran Kumar Bhatt, Sanjay Pattanshetty, and Helmut Brand

Abstract

House to around 65 million people, the Small Islands Developing States (SIDS) is a grouping of low-lying states in three geographical regions: the Pacific and Atlantic, the Indian Ocean and South China Sea, and the Caribbean. The SIDS population primarily depends on natural resources for income due to their geographic settings. These also serve as a source of food supply, water, and livelihood for the communities while holding aesthetic and spiritual value. This chapter examines climate challenges faced by SIDS (including public health risks) and their role in the climate change negotiations. It highlights the need to identify mechanisms and provide advisory support and financing to boost the resilience-building efforts of these islands, which could be effectively initiated through diplomacy and negotiations at the global level.

Keywords

Small islands developing states · Climate change · United Nations Framework Convention on Climate Change · Paris Agreement

---

K. K. Bhatt (✉)
Centre for Health Diplomacy, Department of Global Health Governance, Prasanna School of Public Health, Manipal Academy of Higher Education (MAHE), Manipal, India

S. Pattanshetty · H. Brand
Department of Global Health Governance, Prasanna School of Public Health, Manipal Academy of Higher Education (MAHE), Manipal, India

H. Brand et al. (eds.), *Climate Change and Public Health*,
https://doi.org/10.1007/978-981-95-0895-2_12

12.1 Introduction

House to around 65 million people, the Small Islands Developing States (SIDS) refers to a grouping of low-lying states in three geographical regions: the Pacific and Atlantic, the Indian Ocean and South China Sea, and the Caribbean. The SIDS form a distinct grouping consisting of 38 states and 20 Associate Members of the United Nations (About small island developing states n.d.). One of the exclusive features of the SIDS is the ocean area, which is under their control—Exclusive Economic Zones (EEZ), which according to the United Nations, is more than 28 times their landmass. Despite this, the relatively small population, remoteness from major markets, and volatility of marine ecosystems pose a host of challenges to the effective development of SIDS. Since most SIDS are remotely located, they incur high import and export costs (Pelling and Uitto 2001). The SIDS population primarily depends on natural resources for income due to their geographic settings. These also serve as a source of food supply, water, and livelihood for the communities. Additionally, it holds aesthetic and spiritual value.

The UNFCCC has highlighted that SIDS are at risk of the impacts of climate change, such as rising sea levels and erosion of coral reefs. The vulnerability of the SIDS to natural hazards makes them susceptible to economic and health shocks beyond domestic control.

The frequency of major natural disasters annually was recorded around ten before the 2000s. However, in the first two decades of the twenty-first century, the frequency has increased to 20 each year, as shown in Fig. 12.1. The World Bank Group found that approximately USD 12.6 billion worth of damage would be incurred in the Caribbean region due to climate-related hazards (Rozenberg et al. 2020). Thus, there is a need to mobilize funds and resources to enhance the resilience of SIDS towards the rising frequency and intensity of natural disasters.

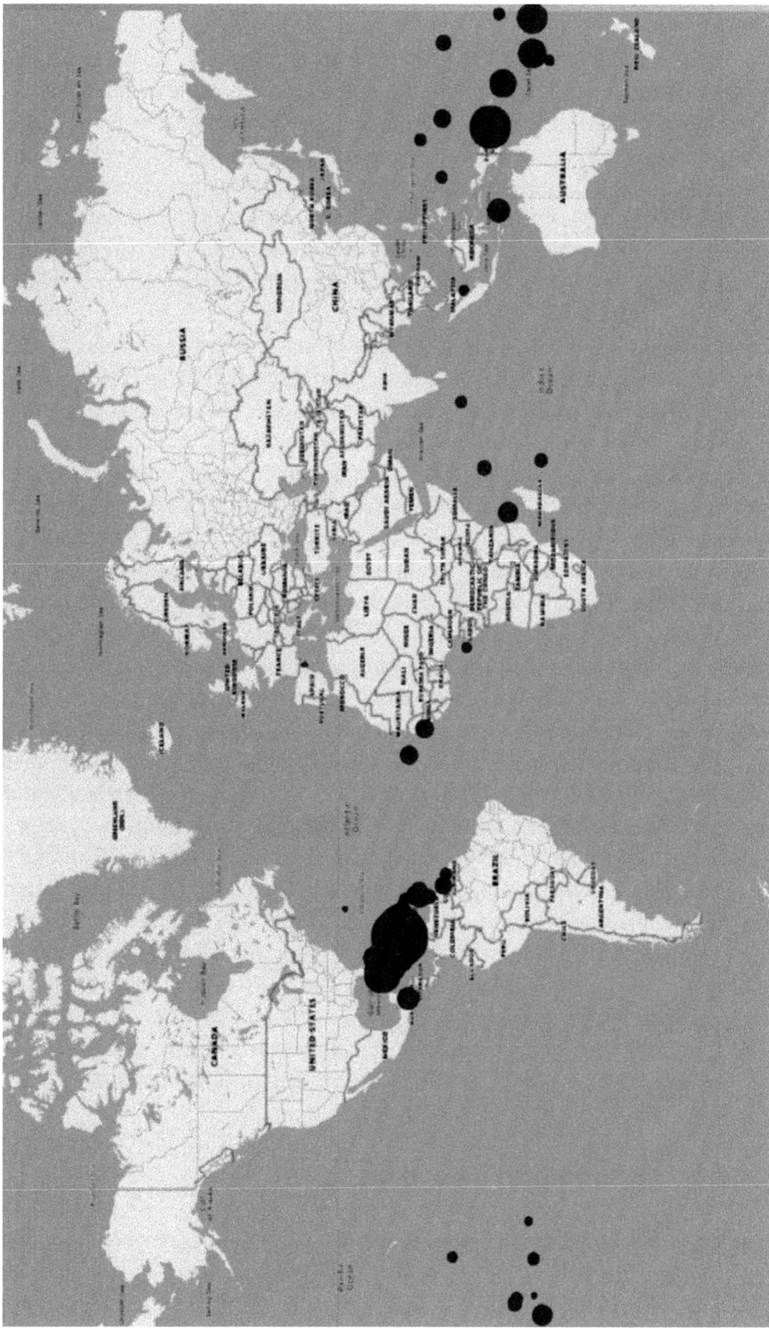

Fig. 12.1 Incidence of Natural Disasters in the SIDS post-2000. (Source: EM-DAT The International Disaster Database)

12.2 Evolution of Debates on Climate Change in the Global Diplomacy Landscape

Over the years, climate change has grown to become one of the most discussed issues in international negotiations. Often regarded as a complex policy issue, the adaptation and mitigation measures to counter climate change can be understood through its temporal and conceptual dimensions. The debate begins with the question of who is responsible for the problem, as most developing countries, who are primarily held responsible for harmful emissions, opine that the industrial economies should be responsible for their historical emissions. The damage already caused, mixed with the current emission, could impact future generations, with the poor and minor contributors to the issue suffering the most.

On the conceptual dimension, climate change must be viewed globally by focusing on its causes and effects. Such an outlook would highlight the need for collective action and address the root causes through enhanced international cooperation. International Relations (IR) Theories, such as constructivism, can be used to understand the paradigm shift in climate change-related negotiations and discussions in global politics (Okereke 2010). The constructivists emphasize the shifting away from rationalist and interest-based factors to the role of the creation of knowledge and shaping the position of nation-states. According to them, cooperation among nations is influenced by discursive power and ideational elements and is not limited to material or power factors (Muni 2020). Further, while other IR theories, such as neoliberal institutionalism, view relations through a set of structural parameters that are established by the existing system, constructivism allows for a dynamic structure. As per the theory, even though states lean toward power-obtaining prospects, non-state actors and newer norms influence the state's selection of a realist path. The non-state actors include transnational organizations, multilateral forums, and even individuals. In terms of Climate Diplomacy, the IPCC is an example, which through its scientific assessments, has been shaping the climate agenda (*About the* IPCC 2023). The formation of the UNFCCC, the signing of the Kyoto Protocol, and the Paris Declaration highlight the facet of addressing climate change through interstate governance. Constructivism offers a paradigm to explain the attempts in global governance to include climate change by including non-state and transnational actors. The constructivist lens allows for understanding climate change as a social process and explaining the formation of concepts like sustainable development and responsibility due to historical emissions, which the SIDS centre their argument around.

12.3 Impact of Climate Change on SIDS

The previous section attempted to analyse the inclusion of climate change in global politics and how transnational actors play a role in shaping the agenda. SIDS are among the most minor contributors towards climate change but are often cited as the most affected. For instance, the total GHG emissions by the SIDS account for less than 1%, as shown in Graph 12.1. Due to their unique situation, SIDS have been

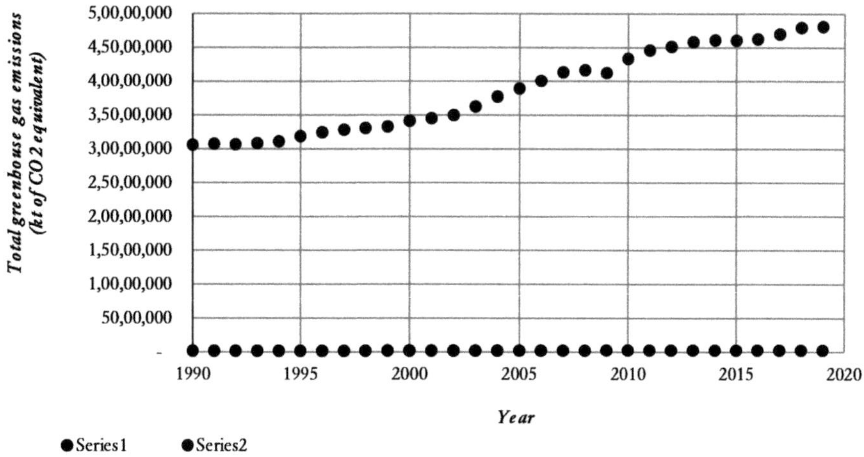

Graph 12.1 Total greenhouse gas emissions—SIDS and the World. (Source: World Bank 2023)

acknowledged during international negotiations with Governments agreeing on the need for coordinated actions. While climate change is expected to affect a global scale, SIDS are more vulnerable due to its geographic settings.

The rise in sea levels impacts the availability of freshwater resources due to saltwater intrusion into aquifers. Some SIDS, such as the Bahamas, Barbados, and Mauritius, depend on groundwater and would face issues finding a supply for domestic usage. Studies have proven that food security and production are directly related to climate change; the concern is grievous for SIDS, which depends on imports and subsistence farming (Climate Change Secretariat 2005). While the increasing saltwater intrusion stresses agricultural production, the changes in weather patterns, such as an increase in heat, tropical cyclones, and droughts, change the soil moisture levels. This, in turn, impacts subsistence farming. Further, the increasing water level also leads to a reduction in the land available for agriculture. The populations of SIDS draw a significant source of their protein from the fisheries, which is also a mode of employment. The availability of fish is being affected due to increasing water temperature, which is forcing the people in SIDS to look for alternative sources (Monnereau and Oxenford 2017).

A final aspect of climate change is its impact on human health in the SIDS. Climate change acts as a risk multiplier due to its impact on social, economic, and environmental determinants of health (Tukuitonga and Vivili 2021). Most SIDS lie within the tropical zone, making them more suitable for transmitting tropical diseases such as malaria, filariasis, and dengue. The erratic weather patterns have furthered the incidence of water and VBDs. Additionally, exposure to increased extreme weather events increases the threat of physical injuries and loss of property. Since many critical infrastructures, such as airstrips and hospitals in the SIDS, are found along the coasts, they are prone to climate-induced structural and functional pressures, which affect their capacity to provide services (Pan American Health Organization 2018). Thereby affecting accessibility to healthcare and highlighting the need for climate and disaster-resilient infrastructure development.

12.4 Locating the SIDS in Climate Change Negotiations

SIDS, which account for the least responsible for climate change, are touted to be some of the worst affected and suffer from its impacts. From losing arable lands and valuable resources, making habitation more challenging to complete submersion forcing evacuation, the SIDS thus make a particular case that demands the immediate attention of the international community. The countries might not be able to voice their concern at the individual level, but as a grouping, the SIDS have been able to influence climate change negotiations. The issue of climate change has been a part of international negotiations since the 1992 United Nations Conference on Environment and Development, held in Rio de Janeiro (Fig. 12.2).

However, in most international negotiation settings, structural power continues to be a deciding factor. This is backed by the fact that most of the discourses in these discussions were primarily developed by industrialized economies and blocks such as the EU, the USA, and emerging markets like China. However, with significantly less structural power and presence, it is more difficult for SIDS to influence policy outcomes in negotiations. Thus, SIDS have ever since been involved in creating a collective new discourse centred on their extreme vulnerability, devastatingly impacting ecosystems and livelihoods, despite being among the smallest contributors to climate change. The SIDS have also resorted to measures that have placed them in the global diplomacy arena for shaping the climate agenda. They have been the forerunners in ratifying climate action plans such as the United Nations Convention on Climate Change or the Kyoto Protocol. Similarly, in 2009, Maldives held what is now called the first underwater Cabinet meeting. Former Maldivian President Mohamed Nasheed explained the event as a message to the world about what is happening and what would happen if no concrete climate actions were taken (Rasheed 2019). More recently, the Foreign Minister of Tuvalu recorded his speech during the COP26 in Glasgow in knee-deep seawater to highlight the plight of his nation due to climate change (Reuters. 2021). The definition of climate change by the SIDS has contrasted mainly with that of other influential players, such as the United States or China, who view it as an economic and energy issue. Although unable to attain a discursive hegemony on their discourse, the island states, through effectively participating in diplomatic negotiations, have built it as an existential issue with the survival of millions at stake.

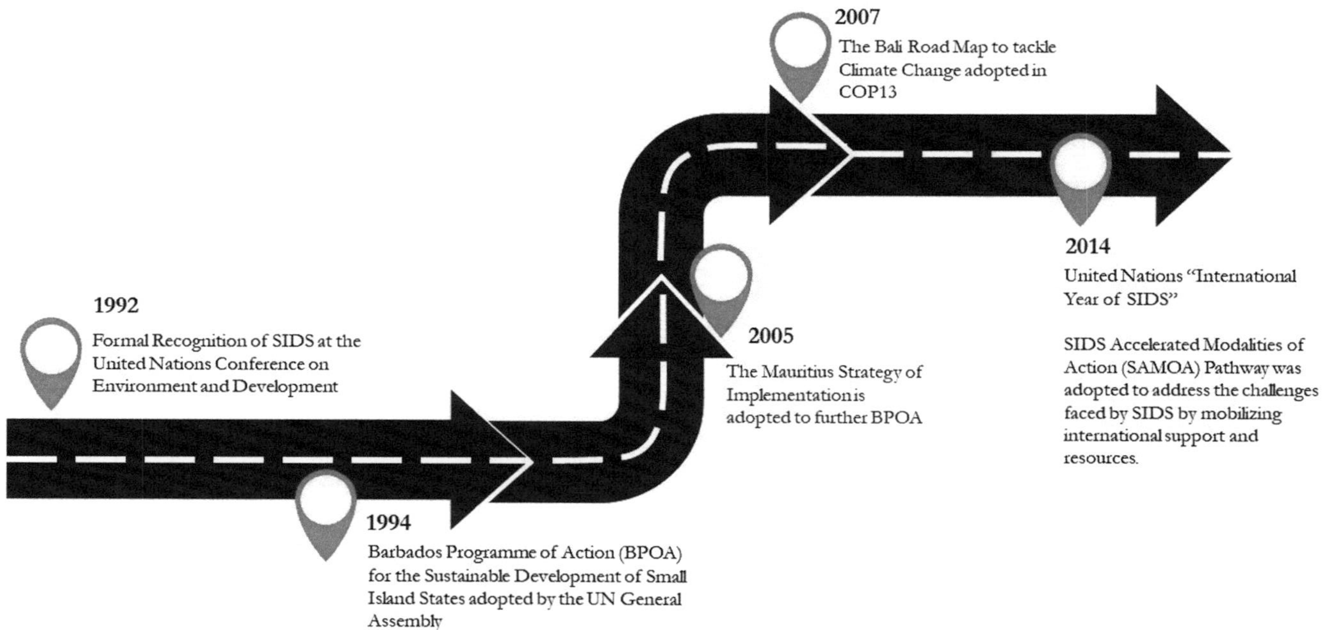

Fig. 12.2 Milestone decisions at the global level concerning the SIDS. (Source: Author's creation based on The Samoa Pathway (UN-OHRLLS) United Nations. (n.d.). The Samoa Pathway. Office of the High Representative for the Least Developed Countries, Landlocked Developing Countries and Small Island Developing States. https://www.un.org/ohrlls/content/samoa-pathway)

12.4.1 Enabling SIDS' Capacity to Adapt and Mitigate Climate Change

The wide range of uncertainties associated with climate change makes it difficult for the SIDS to develop a mitigation mechanism with their limited resources. The SIDS require a mix of policy solutions that cater to the immediate need for recovery from shocks (COVID-19) and, in the long run, enable them to achieve progress as diversified and resilient economies. The process should focus on improving the socioeconomic concerns of SIDS, such as poverty alleviation, employment, and improving educational and health infrastructures. Thus, the threat of variability and the need to adapt to climate change demands changes in SIDS' social and societal norms.

The COVID-19 pandemic exposed the fragile nature of the economies worldwide by affecting all sectors of the economy. Tourism, which is one of the vital sources of income for SIDS too, was severely affected, depriving revenue generation. Therefore, it is necessary to ensure a steady flow of climate finance and funnel it effectively towards enhancing resilience against intensifying natural disasters and climate change. Despite being discussed regularly, access to climate finance scenarios remains relatively less for the SIDS. The COP15 of the UNFCCC held in 2009 witnessed the developed countries committing to mobilizing USD100 billion collectively for climate actions in the developing countries (Thomas and Theokritoff 2021). However, this was extended to 2025 during the COP21 held in Pairs. Within the pledged amount, access to SIDS has remained low. Table 12.1 shows the climate

Table 12.1 Climate finance provided and mobilized (USD billion)

Financing type/year	2013	2014	2015	2016	2017	2018	2019
Bilateral public climate finance	22.5	23.1	25.9	28	27	32	28.8
Multilateral public climate finance attributable to developed countries	15.5	20.4	16.2	18.9	27.5	29.6	34.1
Multilateral development banks	13	18	14.4	15.7	24.1	25.8	30
Multilateral climate funds	2.2	2	1.4	2.6	2.9	3.5	3.8
Inflows to multilateral institutions	0.3	0.4	0.4	0.6	0.5	0.3	0.3
Climate-related officially supported export credits	1.6	1.6	2.5	1.5	2.1	2.1	2.6
Private climate finance mobilised	12.8	16.7	N/A	10.1	14.5	14.6	14
By bilateral public climate finance	6.5	8.1	N/A	5	3.7	3.8	5.6
By multilateral public climate finance attributable to developed countries	6.2	8.6	N/A	5.1	10.8	10.8	8.4
Total	*52.2*	*61.8*	*60.8*	*58.6*	*71.2*	*78.3*	*79.6*

Source: (OECD 2023). Last accessed on 26 July 2023

funds that the various public and private sources have allocated to developing countries. While bilateral financing has been the principal source, the Multilateral Development Banks have emerged as one of the largest financers to the climate funds. As highlighted in it, around USD79.6 billion was ably mobilized for overall climate actions. However, within this, the SIDS were allocated an average of one billion, which they have claimed to be insignificant relative to the risk faced (Climate Funds Update 2021).

Furthermore, the climate funds for SIDS should focus extensively on enhancing their readiness towards adverse effects. The readiness support includes helping with the planning of adaptation measures, focusing on the strengthening of the institutions and guidance on developing strategic frameworks. Of the 58 states and associated members of the UN constituting the SIDS, around 41 of them have received readiness financing, as shown in Graph 12.2.

The SIDS have repeatedly highlighted the absence of a specific window for them under the Green Climate Fund. Expansion of financial support by building a blend of public and private financing is necessary to foster these economies' resilience. While constant discussion exists about the need to act immediately, the work on the ground is primarily limited. While analysing the negotiations and decisions about climate change, it is found that SIDS as a group has had an impact rather than individually. Hence, identifying mechanisms and providing advisory and finance help boost the resilience-building efforts of these islands, which could be effectively initiated through diplomacy and negotiations at the global level through platforms like SIDS. Building partnerships with multilateral forums such as the G20 could help the SIDS ensure that their concerns are considered in the decision-making process. Further, the Multilateral Development Banks could collaborate with partnerships such as the Coalition for Disaster Resilient Infrastructure (CDRI), thereby increasing the availability and accessibility of technical, technological, and institutional assistance for the SIDS to develop climate and disaster-resilient infrastructure. With the impacts of climate change becoming visibly intense, there is a need to raise awareness about the vulnerable situation of SIDS.

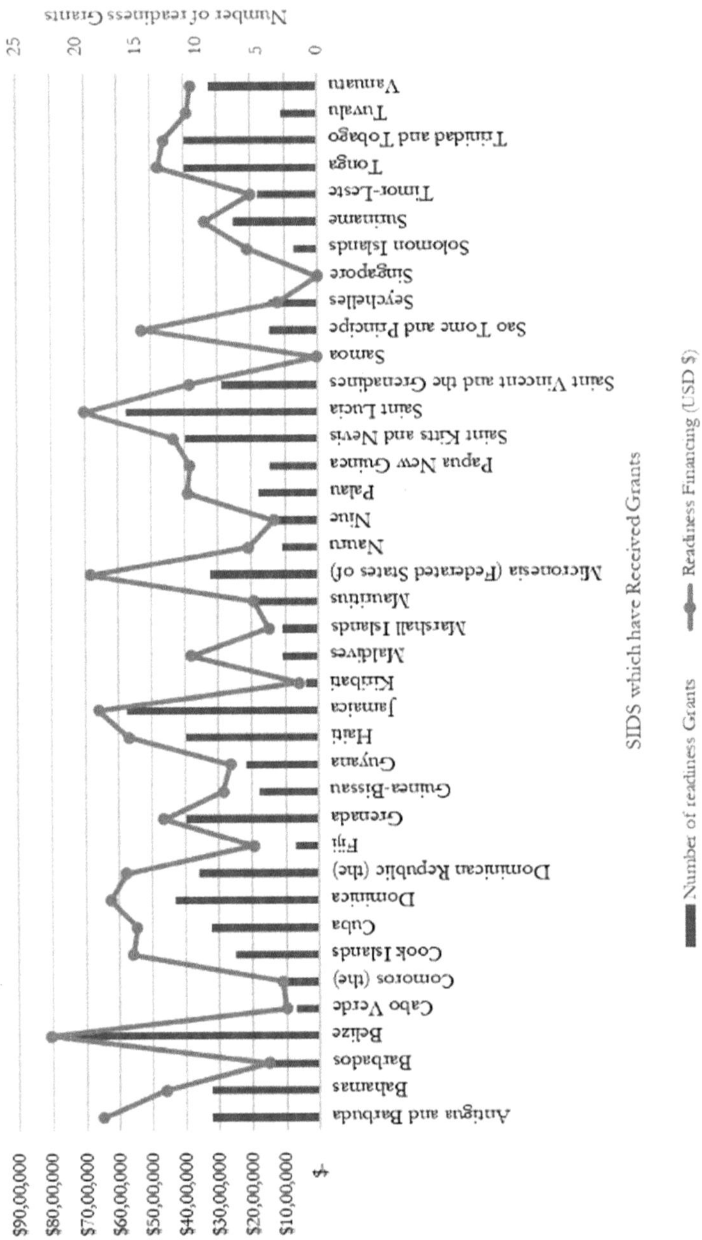

Graph 12.2 Climate readiness financing to the SIDS. (Source: Climate Funds Update 2023)

References

About Small Island Developing States. Office of the high representative for the least developed countries, landlocked developing countries and small island developing states. n.d.. https://shorturl.at/cqxG0

Climate Change Secretariat. Climate change: Small Island developing states. In: UNFCCC. The United Nations Framework Convention on Climate Change; 2005.

Climate Funds Update. Climate finance regional briefing: Small Island developing states (2020) – Climate Funds Update. Climate Funds Update; 2021. https://climatefundsupdate.org/publications/climate-finance-regional-briefing-small-island-developing-states-2020/

Climate Funds Update. Data Dashboard - Climate Funds Update. Climate Funds Update; 2023. https://climatefundsupdate.org/data-dashboard/#1541245745457-d3cda887-f010

ipcc. About the IPCC. The Intergovernmental Panel on Climate Change; 2023. https://www.ipcc.ch/about/

Monnereau I, Oxenford H. Impacts of climate change on fisheries in the coastal and marine environments of Caribbean Small Island…. ResearchGate. 2017; https://shorturl.at/dCLW6

Muni V. Climate change politics through the Constructivist approach. Nepal: NIICE; 2020. https://niice.org.np/archives/6365

OECD. Climate finance provided and mobilised by developed countries in 2013–2021: aggregate trends and opportunities for scaling up adaptation and mobilised private finance. Paris: OECD; 2023.

Okereke C. Climate justice and the international regime. Wiley Interdiscip Rev Clim Chang. 2010;1(3):462–74. https://doi.org/10.1002/wcc.52.

Pan American Health Organization. Climate change and health in Small Island developing states. World Health Organization; 2018.

Pelling M, Uitto JI. Small Island developing states: natural disaster vulnerability and global change. Global Environ Change B Environ Hazard. 2001;3(2):49–62. https://doi.org/10.1016/s1464-2867(01)00018-3.

Rasheed AA. Small Island climate diplomacy in the Maldives and beyond. E-International Relations; 2019. https://www.e-ir.info/2019/06/16/small-island-climate-diplomacy-in-the-maldives-and-beyond/

Reuters. Tuvalu minister to address Cop26 knee deep in water to highlight climate crisis and sea level rise. The Guardian. 2021; Retrieved June 12, 2023, from https://www.theguardian.com/environment/2021/nov/08/tuvalu-minister-to-address-cop26-knee-deep-in-seawater-to-highlight-climate-crisis

Rozenberg J, Browne N, De Vries Robbé S, Kappes M, Lee W, Prasad A. 360° resilience: a guide to prepare the Caribbean for a new generation of shocks. In: Global facility for disaster reduction and recovery (GFDRR). The World Bank; 2020.

Thomas A, Theokritoff E. Debt-for-climate swaps for small islands. Nat Clim Chang. 2021;11(11):889–91. https://doi.org/10.1038/s41558-021-01194-4.

Tukuitonga C, Vivili P. Climate effects on health in Small Islands developing states. Lancet Planet Health. 2021;5(2):e69–70. https://doi.org/10.1016/s2542-5196(21)00004-8.

United Nations. The Samoa pathway. office of the high representative for the least developed countries, landlocked developing countries and small island developing states. n.d.. https://www.un.org/ohrlls/content/samoa-pathway

World Bank. (2023). CO_2 emissions (kt) (Indicator: EN.ATM.CO_2E.KT) [Metadata glossary entry]. Retrieved August 13, 2025, from https://databank.worldbank.org/metadataglossary/millennium-developmentgoals/series/EN.ATM.CO2E.KT

Towards Low-Carbon and Resilient Healthcare Systems

13

Julien Venne and Juliane Mirow

Abstract

Climate change is both a threat to and a consequence of the health system's activities. It affects global health and puts pressure on health systems, which, in turn, account for more than 4% of greenhouse gas emissions, significantly contributing to global warming. A call for action is needed to create future health systems that are both resilient and have a low carbon footprint. While some global initiatives have recently been engaged at the worldwide level, different strategies must be agreed upon, and stakeholders in the healthcare sector need to be widely involved in these new approaches. This chapter presents the most recent findings on the health system's contribution to climate change, highlights evidence-based methodologies for footprint assessment, and suggests possible solutions and ways forward.

Keywords

Low-carbon health systems · Resilience · System thinking · Health systems footprint · Paris Agreement

J. Venne (✉)
Department of Social and Health Innovation, Prasanna School of Public Health, Manipal Academy of Higher Education (MAHE), Manipal, India
e-mail: julien.venne@malsamamind.com

J. Mirow
European University Alliance CHARM-EU, University of Barcelona, Barcelona, Spain

H. Brand et al. (eds.), *Climate Change and Public Health*,
https://doi.org/10.1007/978-981-95-0895-2_13

13.1 Introduction

Climate change is both a threat and a consequence of the health system's activities. Indeed, climate change is affecting global health and putting pressure on health systems, which in turn are accounting for more than 4% of GHG emissions, which contribute heavily to global warming. A call for action is needed to create future health systems, both resilient and with a low carbon footprint. Some initiatives have been recently engaged at the worldwide level, but different strategies must be agreed, and the healthcare sector's stakeholders need to get involved widely in these new approaches. This chapter presents the most recent findings around the contribution of health systems on the climate change phenomenon, highlights evidence-based methodologies for the footprint assessment, and suggests a few solutions and ways forward.

13.2 The Climate Change Challenge

The main challenge of the twenty-first century is certainly the capacity of humanity to thrive and develop its activities while preserving the Earth system. In the last decade, researchers (Steffen et al. 2015) from the Stockholm Resilience Centre have defined a set of nine planetary boundaries within which future generations could develop a good quality of life and wellbeing: climate change, the freshwater change, the stratospheric ozone depletion, the atmospheric aerosol loading, the ocean acidification, the biogeochemical flows, the novel entities, the land-system change, and the biosphere integrity. According to recent scientific work, at least six of these boundaries have already been crossed. For this chapter, we will focus on one of these limits: climate change.

Created in 1988 by the United Nations Environment Program (UN Environment) and the World Meteorological Organization (WMO), the IPCC gathers thousands of scientists across the world to review periodically the state of the research and to publish reports summarizing the main findings around the evolution of Earth's climate. The most recent reports (IPCC 2023a, b) have brought to the world community clear conclusions about the status of climate change and the associated global warming, jointly with a detailed description of its effects as well as a series of objectives and ways forward to mitigate these consequences (IPCC 2023a, b).

The first major statement from the IPCC is the fact that human activities "have unequivocally caused global warming" (IPCC n.d.). The anthropic origin of climate change is demonstrated. Based on this unquestionable identification of causing factors, the IPCC has defined several scenarios regarding the development of the global warming phenomenon and its consequences. The way that humans extract resources, produce and consume goods and services, and organize their activities has generated—and will generate—a negative impact on nature and living beings. Long ago, the relationship between the health of humans and nature was demonstrated (Paracelsus during the sixteenth century). Since then, a large corpus of scientific research has been developed with, for instance, the creation of the environmental health discipline in the second half of the twentieth century and the rise of the concept of One Health, promoted by the WHO for a few decades. As an example of these relations and impact, the WHO reports that air pollution is responsible for more than seven million deaths worldwide and generates a variety of health issues such as respiratory diseases, cardiovascular damage, fatigue, headaches, anxiety, irritation of the eyes, nose, and throat, damage to reproductive organs, nervous system damage (IPCC 2023a). On the other side, the consequences of human activities on climate are nowadays more and more tangible for several populations across the globe facing extreme climate events, severe heat waves, floods, droughts, or heavy storms.

The second strong piece of information from these IPCC reports is that climate change itself is a threat to human wellbeing and planetary health. According to the IPCC's projections (IPCC 2023a), the rise of the temperature is threatening the life of animal species and seagrasses in several parts of the world, especially around the equator where 80–100% would be affected in the worst-case scenarios. Regarding human health, the combination of heat and high humidity levels generates high mortality risk in equatorial zones most days every year. Food production is also strongly impacted whether we are observing the projections related to crops or fisheries.

The third important piece of information to consider is that climate change and global warming are intimately related to the concentration of GHG in the atmosphere. As GHG emissions are directly correlated with human activities, the IPCC advises in its two latest reports to drastically decrease these emissions before 2025 to avoid reaching tipping points in the climate crisis, which could lead to no-return situations having dramatic consequences on life on Earth.

The UN organizes every year the Conference of Partners (COP), gathering all UN member states to discuss the challenge of climate change and negotiate common strategies and objectives between worldwide countries. In 2015, the COP21 held in Paris concluded with an agreement, the "Paris Agreement", engaging each country to reduce its GHG emissions to maintain certain stability of the global climate and maintain global warming under 1, 5° Celsius compared to the pre-industrial era. The UN estimated in 2019 that the United Nations' commitment to this agreement would mean reducing global GHG emissions by 7.6% per year (UNEP 2019). To illustrate the potential impact of such reduction, during the COVID-19 pandemic, because of lockdown measures in most countries, global emissions in 2020 have declined by −6, 4% (Tollefson 2021). The effort required by the Paris Agreement is thus extremely ambitious and requires drastic changes in our human activities' design.

13.3 The Health Systems Footprint and Their Resilience

Therefore, one of the challenges is to engage in a reduction of GHG emissions in the different sectors of human activities. Spontaneously, the population tends to identify heavy industrial sectors, such as the steel or chemical industry or the transport sector planes, trucks, boats as being the most emitting domains. Nevertheless, recent studies (Wang et al. 2020) showed that the healthcare systems in the world contribute to 4, 4% of global GHG emissions, which account for two gigatons of carbon dioxide (CO_2) equivalent. This would represent the total emissions of 514 coal plants, and if all healthcare systems were all together in a country, it would be the fifth emitter on Earth. The healthcare systems are at the same time an infrastructure that allows humanity to manage the effects of climate change on human health and a strong contributor to the problem. Moreover, healthcare professionals have a specific role in society and have created close links with the population. Healthcare workers are trusted and respected by citizens and can have a certain influence on the public and lead by example. This situation has driven the policy- and decision-makers worldwide to engage, during the COP26 organized in Glasgow (UK) in 2021, a large action in favour of low-carbon and resilient health systems: the Alliance for Transformative Action on Climate and Health (WHO 2023). This initiative aims to "build climate-resilient and sustainable health systems" (WHO 2023). The WHO, which is leading and coordinating it, gathers UN member states to engage in knowledge sharing and actions around four working groups on climate-resilient health systems, low-carbon sustainable health systems, supply chains, and financing. To enable it, a community of practice (WHO and Health Security and Environment Cluster 2015) has been launched in June 2023.

The concept of resilience has been defined by the IPCC as the "capacity of a social-ecological system to cope with a hazardous event or disturbance, responding or reorganizing in ways that maintain its essential function, identity, and structure, while also maintaining the capacity for adaptation, learning, and transformation" (IPCC 2014). The WHO has published an operational framework (WHO and Health Security and Environment Cluster 2015) including six key building blocks: leadership and governance, health workforce, health information systems, essential medical products and technologies, service delivery, and financing. This document aims to guide public health practitioners, but also a decision or policymakers, to "enhance the capacity of health systems to protect and improve population health in an unstable and changing climate" (WHO and Health Security and Environment Cluster 2015). It also defined a climate-resilient health system as "one that is capable of anticipating, responding to, coping with, recovering from and adapting to climate-related shocks and stress, to bring sustained improvements in population health, despite an unstable climate" (WHO and Health Security and Environment Cluster 2015).

Two instruments can be mentioned as central to this approach. First, the health systems must assess their exposure and the potential impact that a changing climate could have on them. The WHO has proposed the "vulnerability and adaptation (V&A) assessment" (WHO and Health Security and Environment Cluster 2015). The level of risks will depend on three factors: the exposure of a specific area depending on its physical location and geography and the projected climate disruptions that may occur; the characteristics of the population, including its health profile and the associated determinants of health (social, political, commercial, environmental, etc.); and the capacities and resources of the health system facing such disruptions. Additionally, the WHO has published several recommendations for the definition and the establishment of the "Health National Adaptation Plan" (HNAP) (World Health Organization and International Institute for Sustainable Development 2021). This instrument supports decision-makers in the planification of actions and initiatives aiming to increase their resilience capacities, e.g., to cope and adapt more effectively to climate change and other environmental adverse effects, considering the vulnerability level (through the V&A assessment) on several dimensions.

13.4 Evaluating GHG Emissions of Health Systems

From the national adaptation plans to micro-level healthcare settings, the contribution of healthcare systems to the global GHG emissions, and so to the global warming phenomenon, drives the health systems managers to look for solutions to the decarbonization of their activities. The first step in this approach is to evaluate the specific emissions of healthcare services, both through the identification of their sources and their respective weight in the total emissions of the system. The main emissions accounting framework and methodology is the GHG Protocol (GHG Protocol), an initiative of the World Resources Institute and World Business Council for Sustainable Development that provides globally standardized frameworks for measuring and managing climate-warming emissions within the private and public sectors (GHG Protocol 2004, 2011, 2015) (EPA n.d.). With this instrument, we classify GHG emissions into three categories called "scope". Scope 1 accounts for the direct emissions, which occur from sources directly owned or controlled by the organization, typically stationary combustion like furnaces, vehicles, or machinery generating emissions. Scope 2 accounts for electricity-related indirect GHG emissions from the generation of purchased electricity consumed by the organization (they are qualified as indirect emissions as they occur at the level of the electricity provider but are driven by the organization's consumption). Scope 3 accounts for all other indirect GHG emissions generated because of the activity of the organization but from sources that are not directly owned by it (typically the use of purchased products or services whose lifecycle is generating emissions). The application of this framework to the healthcare sector can help managers and decision-makers assess their emissions around these three scopes:

- Scope 1 includes mainly the emissions generated by the combustion systems or waste management systems owned by a healthcare system or a healthcare facility, its vehicles, as well as the fugitive emissions from the cooling systems or medical usage (such as anaesthetic gases).
- Scope 2 covers the emissions related to purchased electricity and other streams around heat and cooling systems.
- Scope 3 includes the emissions related to medical equipment and instruments, pharmaceutical products, food and catering services, products, employees, and

patients commuting and travelling, construction, inhalers, waste off-site management, or any other procured goods or services.

Recent studies (Health Care Without Harm and Arup 2022) have analysed at the global level the emissions of healthcare systems and show that Scope 1 accounts for 17, 5% of the total emissions, Scope 2 for 12, 6%, and Scope 3 for 69, 9%. If we combine the three scopes, the share of GHG emissions related to energy provision and consumption would represent more than 50% of the total GHG emissions of healthcare systems. Thus, energy systems' transformation is a major stake in this dynamic (Fig. 13.1).

The urgent action supported by the United Nations and WHO with the ATACH initiative (WHO 2023) is the first to identify the sources of GHG emissions and to quantify them, both at the national level for the whole healthcare systems or at the micro-level for healthcare settings. Three different methods (Health Care Without Harm and Arup 2022) could be employed and combined to obtain more accurate results. The "Top-down approach" combines the evaluation of expenditures of an organization around the identified GHG emissions sources, with existing readily available international or national databases, like the Environmentally Extended Input-Output (EEIO) databases (Kitzes 2013), which provide references about the estimation of emissions based on monetary terms. The benefit of this approach lies in its comprehensiveness and its ability to prevent any leakage as it covers all the hidden expenses including those related to operational and supply chain. It is also interesting in a context where data points can be missing, or doubts could exist about their accuracy. The challenge is that it might be difficult to analyse in a fine and granular approach the emissions of specific settings. Another method is called the "Bottom-up approach", which will study the activities of the organization, a hospital for instance, from cradle to the grave, also called Life-Cycle Assessment (LCA) as defined by the International Organization for Standardization (ISO 2022), of the products and services involved in the different activities. For a specific product, the consumption of resources from the raw material extraction, through the manufacturing process, the distribution, the use, and then the waste management and recycling processes are considered. The consumption of resources and associated actions to develop this specific product or the whole activities are then combined with emissions factors defined by sectors and types of activities by official agencies such as the United States Environmental Protection Agency (Kent et al. 2016). The advantage of the bottom-up approach is that it allows more granularity in the analysis, specifically for studies at the micro-level. The drawbacks are related to the potential lack of accurate data to quantify the consumption or hidden aspects of the lifecycle of products and services. A bottom-up approach can be used for the emission sources that are not included in the EEIO databases, for instance, the anaesthetic gases and metered dose inhalers. Lastly, a hybrid approach could be adopted, combining top-down and bottom-up approaches to cover the full spectrum of GHG emissions across the three scopes, depending on the context of the study and the availability of data.

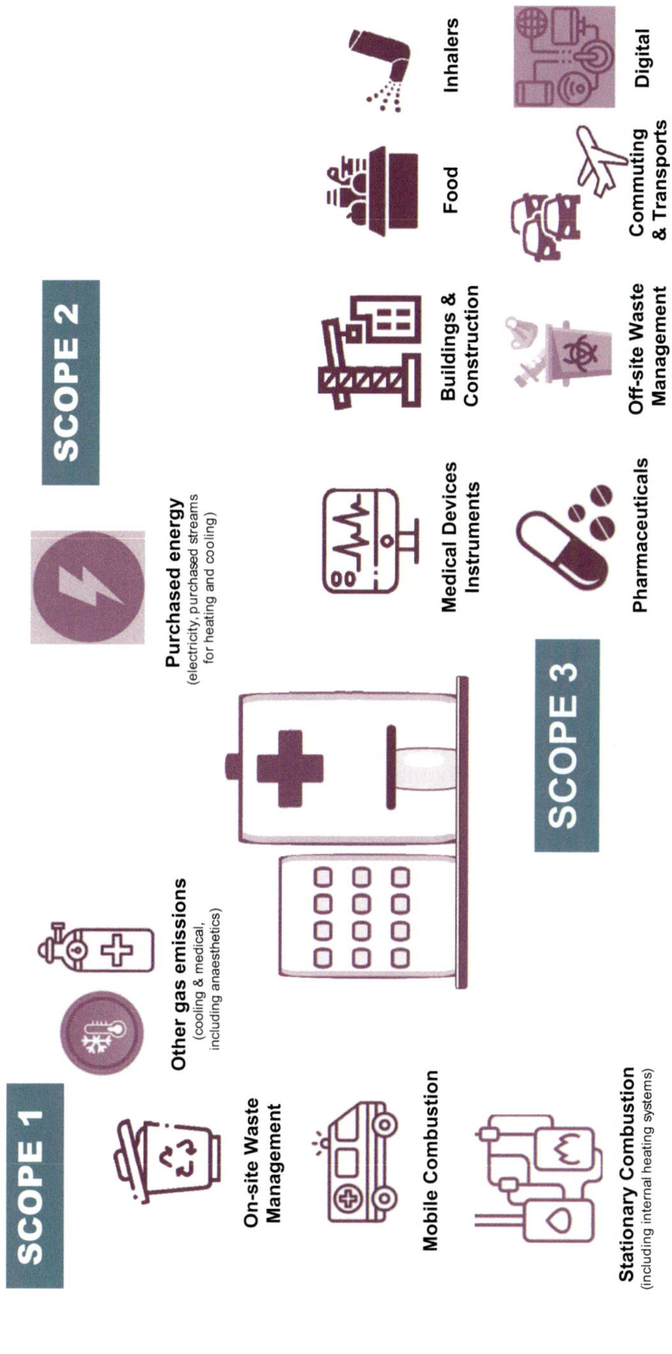

Fig. 13.1 Scopes and main sources of GHG emissions in health systems

13.5 Mitigation Measures to Reduce the GHG Emissions Footprint of the Health Systems

Once the identification of GHG emissions sources and their quantification is done, then different strategies and innovative solutions could be implemented to reduce the emissions. We report here some solutions have produced a positive impact on the footprint of healthcare facilities or the healthcare industry. Around the emissions related to scope 1, health systems could envisage solutions related to their transportation systems such as electric vehicles (Choma et al. 2020), even if a specific assessment of the electricity origin in the specific location of activity will be required to assess the real benefit in terms of reduction of GHG emissions as electricity produced from coal plants will not have the same carbon footprint as nuclear power for instance. Drones (Hiebert et al. 2020) have been experimented on as well for the delivery of medications or biological samples. Regarding anaesthetic gas (Yasny and White 2012), several strategies include the use of less polluting components (from desflurane to sevoflurane or isoflurane) or some changes in the usage processes to avoid gas waste or else using intravenous anaesthesia. Also, more healthcare providers are renovating buildings with more efficient isolation or constructing new premises on the model of net-zero infrastructures (Kaiser 2023), which in turn reduces the use of heating and cooling systems.

Regarding scope 2 and in particular for reducing energy consumption, the strategies could relate to gains in the efficiency of the equipment consuming electricity (Department of Energy Hospital Energy Alliance and US 2011), to a switch in the type of energy used by the organization, through renewable energy (Bank 2017), for instance, or "sufficiency" or "sobriety" (Hu et al. 2023), e.g., the changes leading to the reduction of the energy consumption. Examples of interventions include the use of efficient lighting or heating/cooling systems, the installation of wind turbines or solar panels in healthcare facilities, the development of a Virtual Power Purchasing Agreement (VPPA) (Kent et al. 2016) with renewable energy providers to secure the supply for the needs of the healthcare setting, or else some behavioural interventions towards employees and patients to decrease the use of energy, as switching off equipment and lighting when not used. For the indirect emissions (scope 3), the interventions are always more difficult for an organization to implement as the sources are situated out of the perimeter or control of the healthcare settings. Nevertheless, some initiatives can be launched such as around the transportation of patients and staff with behavioural interventions. Patients and employees could be incentivized to use public transportation or non-emitting mobility solutions such as active mobility also associated with health benefits (Arnott et al. 2014). Alternatives involving organizational changes can be identified to limit transport such as remote work for employees or telemedicine (Purohit et al. 2021) for patients who could receive care and attend a consultation from their home or local points of care. Here again, digital solutions could represent an opportunity to decrease GHG emissions, but a specific assessment is needed as technologies rely on data centres and connectivity networks which consume large volumes of energy themselves. Within the healthcare supply chain, representing the larger part of emissions, several approaches

Few examples of decarbonisation measures for health systems		
Scope 1	**Scope 2**	**Scope 3**
• Transportts: electric vehicles, drones • New forms of anaesthetic gases • Reducing heating/cooling consumption via building isolation	• Efficiency gains on electric equipment • Switch toward revewable energies • Behaviour change interventions to reduce energy consumption • Sufficiency or sobriety	• Digital Health & Telemedicine • Public transports for patients/employees commuting • Re-use, re-purpose or recycle medical equipment • Green chemicals for medicines • Procurement with eco-friendly criteria

Fig. 13.2 Few examples of decarbonization measures for health systems

and solutions could be envisaged. Around medical products, alternative solutions could be engaged to promote the reduction of single-use items, to re-purpose and reuse medical devices and equipment, as well as to develop more recycling processes. Green chemicals (Mishra et al. 2021) are used in the pharmaceutical industry to replace some more contaminating substances. Interesting approaches are focusing on redesigning care practices such as with the "green social prescribing" approach (NHS n.d.) or care organizations to reduce the use of medical equipment or the reduction of appointment frequency.

The development of personalized medicine and targeted care supports a logic of better care and services to the population; at the same time it allows for adjustment of the resources (energy, drugs, devices) to the right need for each patient, contributing also to decarbonization. Similarly, a shift towards more preventive measures and health promotion (Frumkin et al. 2008) would decrease the burden of some diseases, in particular, non-communicable diseases (NCDs) and cancers on healthcare systems and so reducing the related GHG emissions, while improving health, well-being, and quality of life for patients and populations. Around food and catering, hospitals can choose local producers, providing fresh, regional, and seasonal plant-based meals and purchasing organic and certified producers. In general, it could be beneficial for healthcare systems at large (public and private sectors) to engage new practices in the procurement (Ahsan and Rahman 2017) of external goods and services to drive the change across the supply chain. Environmental criteria and suppliers' footprint assessments could be required for their participation in tenders, which could be a way to extend the control of the health systems on this scope 3 (Fig. 13.2).

13.5.1 Adopting a System-Thinking Approach to Reduce the Health Systems' Contribution to Global Warming

In conclusion, health systems decision-makers across the world are becoming more and more aware of both the risk of environmental disruptions such as climate change on human health and healthcare provision as well as the strong responsibility that health systems have in the emission of GHG and so, in the phenomenon affecting our planet Earth and the environment such as the climate change and the global warming. Methodologies have been developed to assess these risks as well as to identify the health systems' emissions sources and to quantify them. Technical or non-technical solutions exist and have proven their impact on the carbon footprint of health systems or facilities. Nevertheless, it seems necessary to develop further communication and education or training actions towards policymakers, health professionals, and healthcare managers to equip them with accurate knowledge and know-how resources. These stakeholders should lead the transformation of the whole healthcare sector and, by example, lead societal changes in front of the main challenge of our century. Processes for standardization and normalization associated with new regulations should create the conditions for the uptake of "green strategies" by healthcare players, from the care provision to the supply chain industry. The complexity of the implementation of solutions can often be related to the required trades-off that they imply and which involve a fine analysis of the impact between different options. Indeed, the deployment of a low-carbon solution could generate additional emissions at another point in the lifecycle of a specific medical process or product. For instance, the development of reusable medical items can reduce their footprint compared to single-use ones but will generate additional new emissions related to the cleaning and sterilization processes. Another emblematic example is around digital health: teleconsultation or remote monitoring solutions could appear as eco-friendly solutions avoiding several non-necessary transports and so fuel-generated emissions, but digital technologies are consuming large quantities of energy and mobilize raw materials (rare metals), which are less and less abundant on our planet. None of the solutions is ideal and some choices will have to be made by decision-makers depending on their local context.

Another important aspect to underline is related to the concept of climate justice. Not all countries are equal in front of climate change. If the IPCC reports have demonstrated that climate change was due to human activities, low- and middle-income countries (LMIC) have contributed less to GHG emissions than industrialized countries, even as they are more exposed to the adverse effects of climate change. Additionally, the LMICs still suffer from a lack of access to healthcare and a lack of resources and have difficulty ensuring Universal Health Coverage (UHC) across their population. These healthcare systems require more investments in the next years, and so we can anticipate that their GHG emissions will increase hand-to-hand with this development. That is why a differentiated agenda should be agreed upon where developed countries engage more drastic efforts around healthcare decarbonization, meanwhile, LMICs engage in the development of their health systems using the most eco-friendly approaches and strategies to build low-carbon and resilient systems.

Finally, the alignment of interests of all stakeholders, namely, policymakers, healthcare providers, health professionals, healthcare supply chain and industry, as well as the patients and the whole population, is a key challenge to creating the condition of a consensus in favour of a trajectory designing both strategies for the resilience of these systems as well as decarbonization approaches to limit the environmental footprint. The recursiveness of both approaches, the climate change affecting human health calling for stronger and larger healthcare systems at the same time of the health systems being more and more responsible for the GHG emissions and global warming, makes this challenge very complex, and contradictory processes must be dealt with. The commitment of the number of United Nations member states to the ATACH initiative led by the WHO is paving the way for the future of green and resilient health systems.

References

Ahsan K, Rahman S. Green public procurement implementation challenges in Australian public healthcare sector. J Clean Prod. 2017;152:181–97. https://doi.org/10.1016/J.JCLEPRO.2017.03.055.

Arnott B, Rehackova L, Errington L, Sniehotta FF, Roberts J, Araujo-Soares V. Efficacy of behavioural interventions for transport behaviour change: systematic review, meta-analysis and intervention coding. Int J Behav Nutr Phys Act. 2014;11(1):1–23. https://doi.org/10.1186/S12966-014-0133-9/TABLES/8.

The World Bank. Climate-smart healthcare: low-carbon and resilience strategies for the Health sector. Washington DC: World Bank Group; 2017. p. 1–80. https://doi.org/10.1596/27809.

Choma EF, Evans JS, Hammitt JK, Gómez-Ibáñez JA, Spengler JD. Assessing the health impacts of electric vehicles through air pollution in the United States. Environ Int. 2020;144:106015. https://doi.org/10.1016/J.ENVINT.2020.106015.

Department of Energy Hospital Energy Alliance, & US. Energy-efficient hospital lighting strategies pay off quickly. US Department of Energy; 2011. https://www1.eere.energy.gov/buildings/publications/pdfs/alliances/hea_lighting_fs.pdf

EPA. GHG emission factors hub. United States Environmental Protection Agency; n.d.. Retrieved July 14, 2023, from https://www.epa.gov/climateleadership/ghg-emission-factors-hub

Frumkin H, McMichael AJ, Hess JJ. Climate change and the Health of the public. Am J Prev Med. 2008;35(5):401–2. https://doi.org/10.1016/j.amepre.2008.08.031.

Health Care Without Harm, & Arup. Designing a net zero roadmap for healthcare: technical methodology and guidance. Health Care Without Harm; 2022. p. 1–73. https://noharm-europe.org/sites/default/files/documents-files/7186/2022-08-HCWH-Europe-Designing-a-net-zero-roadmap-for-healthcare-web.pdf

Hiebert B, Nouvet E, Jeyabalan V, Donelle L. The application of drones in healthcare and health-related Services in North America: a scoping review. Drones. 2020;4(3):30. https://doi.org/10.3390/DRONES4030030.

Hu S, Zhou X, Yan D, Guo F, Hong T, Jiang Y. A systematic review of building energy sufficiency towards energy and climate targets. Renew Sust Energ Rev. 2023;181:113316. https://doi.org/10.1016/J.RSER.2023.113316.

IPCC. II glossary editorial board co-chairs: TSU facilitation. In: Intergovernmental panel on climate change; 2014. p. 1–20. https://www.ipcc.ch/site/assets/uploads/2018/02/WGIIAR5-AnnexII_FINAL.pdf, https://www.ipcc.ch/site/assets/uploads/2018/02/WGIIAR5-AnnexII_FINAL.pdf.

IPCC. Summary for policymakers. In: Romero J, editor. Climate change 2023: synthesis report; 2023a. A Report of the Intergovernmental Panel on Climate Change. https://www.ipcc.ch/report/ar6/syr/downloads/report/IPCC_AR6_SYR_SPM.pdf.

IPCC. Climate change widespread, rapid, and intensifying–IPCC. Intergovernmental Panel on Climate Change; n.d.. Retrieved July 15, 2023b, from https://www.ipcc.ch/2021/08/09/ar6-wg1-20210809-pr/

ISO. IWA 42:2022 – Net zero guidelines. International Organization for Standardization; 2022. https://www.iso.org/contents/data/standard/08/50/85089.html

Kaiser. Kaiser in Santa Rosa achieves net-zero energy | KSRO. Kaiser; 2023. https://www.ksro.com/2023/04/12/kaiser-in-santa-rosa-achieves-net-zero-energy/

Kent C, EPA, GPP. Introduction to virtual power purchase agreements. In: United States environmental protection agency; 2016. p. 1–26. https://www.epa.gov/sites/default/files/2016-09/documents/webinar_kent_20160928.pdf.

Kitzes J. An introduction to environmentally-extended input-output analysis. Resources. 2013;2(4):489–503. https://doi.org/10.3390/RESOURCES2040489.

Mishra M, Sharma M, Dubey R, Kumari P, Ranjan V, Pandey J. Green synthesis interventions of pharmaceutical industries for sustainable development. Curr Res Green Sustain Chem. 2021;4:100174. https://doi.org/10.1016/J.CRGSC.2021.100174.

NHS. NHS England green social prescribing. NHS England; n.d.. Retrieved July 14, 2023, from https://www.england.nhs.uk/personalisedcare/social-prescribing/green-social-prescribing/

Purohit A, Smith J, Hibble A. Does telemedicine reduce the carbon footprint of healthcare? A systematic review. Future Healthc J. 2021;8(1):e85–91. https://doi.org/10.7861/FHJ.2020-0080.

Steffen W, Richardson K, Rockström J, Cornell SE, Fetzer I, Bennett EM, Biggs R, Carpenter SR, De Vries W, De Wit CA, Folke C, Gerten D, Heinke J, Mace GM, Persson LM, Ramanathan V, Reyers B, Sörlin S. Planetary boundaries: guiding human development on a changing planet. Science. 2015;347(6223) https://doi.org/10.1126/SCIENCE.1259855/SUPPL_FILE/STEFFEN-SM.PDF.

Tollefson J. COVID curbed carbon emissions in 2020 – but not by much. Nature. 2021;589(7842):343. https://doi.org/10.1038/D41586-021-00090-3.

UNEP. Cut global emissions by 7.6 percent every year for next decade to meet 1.5 °C Paris target [Press release]. United Nations Environment Programme; 2019. https://www.unep.org/news-and-stories/press-release/cut-global-emissions-76-percent-every-year-next-decade-meet-15degc.

Wang J, Karliner J, Slotterback S, Boyd R, Ashby B, Steele K, Wang J, Weatherup C, Weatherup C, Zambon F. Health care's climate footprint: the health sector contribution and opportunities for action. Eur J Pub Health. 2020;30(Supplement_5) https://doi.org/10.1093/EURPUB/CKAA165.843.

WHO. Alliance for action on climate change and health (ATACH). World Health Organization; 2023. https://www.who.int/initiatives/alliance-for-transformative-action-on-climate-and-health

WHO, & Health Security and Environment Cluster. Operational framework for building climate resilient health systems. World Health Organization; 2015. p. 1–56.

World Health Organization, & International Institute for Sustainable Development. (2021). Review of health in national adaptation plans.. https://www.who.int/publications/i/item/9789240023604

Yasny JS, White J. Environmental implications of anesthetic gases. Anesth Prog. 2012;59(4):154–8. https://doi.org/10.2344/0003-3006-59.4.154.

Mind the Gap Financing the Climate-Health Nexus

14

Uday Khemka and Nitya Mohan Khemka

Abstract

Current climate action pathways predict at least 3 °C of global warming by the end of the century, falling short of the 2015 Paris Agreement's goals. The latest IPCC report indicates that mid- and long-term climate risks are significantly higher than currently observed. There is an urgent need for national adaptation plans with a focus on human health; however, critical investment in resilient health systems remains lacking in climate and health finance discussions. This chapter reviews secondary sources and literature to outline the health financing gap in mitigating climate change effects. It accentuates system changes and initiatives necessary to bridge this gap while presenting actionable recommendations for successful execution.

Keywords

Climate finance · Climate-health nexus · Nationally determined contributions · Health infrastructure · Adaptation · Bridgetown initiative

14.1 The Context

Approximately 3.3 to 3.6 billion people live in contexts that are highly vulnerable to climate change (IPCC 2022). The direct damage costs to health from climate change are estimated to be between USD 2–4 billion annually (WHO 2021). A

U. Khemka
The Nand & Jeet Khemka Foundation, New Delhi, India

N. M. Khemka (✉)
PATH, London, UK
e-mail: nitya@khemka.com

H. Brand et al. (eds.), *Climate Change and Public Health*,
https://doi.org/10.1007/978-981-95-0895-2_14

business-as-usual pathway on climate action would have devastating consequences. By the end of this century, if current national pledges to reduce GHG emissions are implemented, which would still result in at least 3 °C of global warming, significantly below the 2015 Paris Agreement's goal to limit warming to well below 2 °C. According to the latest IPCC report, 'for 127 identified key risks, assessed mid- and long-term impacts are up to multiple times higher than currently observed (*high confidence*)' (IPCC 2022).

According to a 2022 Lancet Report on health and the climate crisis, climate change is the 'greatest global health threat facing the world in the 21st Century' (Romanello et al. 2021). The impacts of climate change on human health are particularly worrisome: climate hazards will contribute to increases in the frequency of communicable and non-communicable diseases; heat is a growing health risk; climate change contributes to food insecurity that leads to malnutrition; and extreme climate events are both direct and indirect drivers of potentially severe impacts on human wellbeing (Romanello et al. 2021).

The UNFCCC process, particularly COP 21 in Paris, has attempted to 'bend the emissions curve' and enshrine ambitious targets. Some of the notable contributions of the Paris COP include:

The shift in emphasis towards national action; cementing of the below 2 °C goal and going further to reference 1.5 °C; the addition of 'global peaking' and balance between emissions and removals as supplementary goals; global stock-taking every 5 years with no backsliding; much more nuanced differentiation between developed and developing countries; greater recognition of adaptation and the inclusion of loss and damage; new developments on finance, including stronger reporting provisions; and renewed recognition of market mechanisms (UNFCCC 2016).

In the wake of the Paris Conference, countries have created Nationally Determined Contributions and increased ambition through updates.

A study suggested that, by 2090, heat waves in Chicago and Paris are expected to occur 25 and 31 percent more frequently, respectively. Additionally, the average duration of a heat wave in Paris is projected to increase from a range of 8–13 days to 11–17 days. Substantial increases in the occurrence of heat waves were also predicted for the western and southern United States and the Mediterranean region (Kinley 2016).

However, this process has had notable limitations recently, including at COP26 and COP27.

The Glasgow Climate Pact, which stated that carbon emissions will need to fall by 45 percent by 2030, barely kept alive the Paris Agreement's target of the 1.5 °C goal. Among the notable limitations of the Glasgow conference were financial support and technical assistance for low- and middle-income countries. There are also clear indications that some of the targets set out at Glasgow have been considerably impacted by the war in Ukraine and ensuing geopolitical challenges that have complicated countries' food and fuel security. Several states have had to reconsider the timelines of their decarbonization efforts to compensate for reduced access to oil and gas supplies. Aside from significant progress on areas such as Loss and Damage (including for a dedicated fund, the governance, and hosting of the Santiago

Network, funding of early warning systems, the Global Shield against Climate Risks program, related funding commitments, and debt relief, etc.), COP27 has been seen not to have delivered anywhere near necessary and time-bound progress on mitigation or adaptation. Given these factors, current policy and technology pathways for climate change mitigation alone will not avert some of the most alarming manifestations of climate impacts. Along with rapid cuts in emissions, the results of which may only start showing by mid-century, there is an urgent imperative to begin work on the design and implementation of national adaptation plans with a focus on human health.

In this context, the critical investment required to establish resilient health systems capable of withstanding climate-related challenges and safeguarding individuals from the adverse effects of climate change is glaringly absent in both the discourses of climate and health financing. This gap in financing, which pertains to the intersection of climate and health, poses a significant risk to the lives of millions, deprives health systems of vital assistance, and exacerbates global health disparities (WHO et al. 2015).

Based upon a review of secondary sources and the existing literature, the paper seeks to outline the landscape of the health financing gap around limiting the effects of climate change. It also points to underlying system changes and initiatives that can help fill the gap and provide reasons to make these amelioration actions more likely to succeed.

14.2 The Challenge

The health sector is confronted with a twofold challenge concerning climate change. It needs to swiftly implement environmentally sustainable practices to reduce its carbon footprint (since the sector activities contribute 4–4.9% of global GHG emissions annually) (Karliner et al. 2019). At the same time, it must rapidly adapt on a large scale to prepare for, minimize, and potentially avoid the adverse impacts that are already affecting human health and health systems (Pichler et al. 2019).

According to data from the Emergency Events Database spanning from 1969 to 2018, there were a total of 10,009 extreme weather events that escalated into disasters, leading to a staggering two million deaths and nearly four million reported cases of diseases (83). The predominant types of extreme weather events on a global scale during this period were floods, accounting for 47 percent, and storms, comprising 30 percent, both of which exhibited an increasing trend (97). When examining the direct weather-related causes of death on a worldwide scale, the highest proportions were attributed to storms at 39 percent, followed by droughts at 34 percent and floods at 16 percent. Notably, drowning emerged as the most prevalent cause of fatality following the onset of flooding (139). Furthermore, it is important to highlight that the health impacts persisted for more than 10 days following a severe flooding event (Ebi et al. 2021a).

Adaptation within the health sector aims to decrease death and illness by enhancing the resilience of health infrastructure to shocks and stresses. This involves utilizing early warning systems for extreme weather events and climate-sensitive infectious diseases, improving disease surveillance, developing plans to address heat-related health risks, and expanding mental health and psychosocial support services, among other measures (Ebi et al. 2021b; Hwong et al. 2022). To achieve these goals, a comprehensive and transformative approach is required, integrating climate risk reduction and building adaptive capacity throughout health systems.

Furthermore, health and climate (and their relationship) should be integrated as cross-cutting factors across all sectors, including food systems, livelihoods, social support structures, water and sanitation, and others. This approach recognizes that the impacts of climate change on health are interconnected with various aspects of society and underscores the need for collaborative action to address these challenges effectively (Cissé et al. 2022).

Following the commitments of the UNFCCC, countries are tasked with developing and regularly updating their NDCs. These NDCs serve to outline the strategies and measures that countries will undertake to achieve their carbon reduction targets as part of their mitigation efforts as well as identify priority areas for adaptation. Among the 185 countries examined, approximately 73 percent (135 countries) of NDCs include references to health. This indicates a recognition of the link between climate change and health outcomes. Around 60 countries specifically mention climate-related health outcomes or adaptation measures, and approximately 40 countries have developed detailed plans for health adaptation (Dasandi et al. 2021).

In a 2021 analysis conducted by the WHO on 19 NAPs, it was found that all the reviewed NAPs identified health as a 'high-priority sector vulnerable to climate change' (WHO and International Institute for Sustainable Development 2021). However, the level of consideration given to addressing the highlighted health risks varied across the plans. To address the intersection of climate change and health, countries are increasingly focusing on the development of specific strategies and plans, such as Health National Adaptation Plans (HNAPs).

Despite these efforts, there exists a disparity between the priorities outlined in written plans and the allocation of financial resources. The WHO's 2021 Health and Climate Change Global Survey revealed that 51 percent of countries reported having a national health and climate change strategy or plan in place (WHO and International Institute for Sustainable Development 2021). However, less than a quarter of the respondents reported achieving high or very high levels of implementation. Insufficient financing or budgetary constraints were identified as key barriers by 70 percent of the countries surveyed, preventing them from fully implementing their strategies and plans to address climate change and health (WHO and International Institute for Sustainable Development 2021).

According to the UNFCCC, it is estimated that global health adaptation will require annual funding ranging from USD 26.8 to 29.4 billion by the year 2050 (Elena et al. 2018). However, many low-income countries struggle to provide basic health services, let alone make the necessary investments to adapt these services to the unfolding risks posed by climate change. The World Bank predicts that by 2050, Sub-Saharan Africa will bear 80 percent of the global health costs resulting from increased cases of malaria and diarrhoeal diseases (World Bank and International

Bank for Reconstruction and Development 2010). Without prioritizing health system adaptation, the growing health burden will not only have significant economic costs but also contribute to substantial non-economic losses and damages associated with climate change.

International climate finance including multilateral development banks (MDBs), International financial institutions (IFIs), and Development financial institutions (DFIs), etc., will play a crucial role, alongside domestic government spending and public-private partnerships and private finance mobilization in bridging the funding gap for health sector adaptation. However, when it comes to climate financing, mitigation efforts receive significantly more public funding compared to adaptation measures across all sectors (OECD 2021). Global tracking indicates that in 2019/2020, all adaptation finance amounted to only USD 46 billion, which is considerably lower than the estimates in the upper range of USD 140 to 300 billion needed for adaptation efforts by 2030 (Buchner et al. 2021). The 2018 Adaptation Gap Report, with a special focus on health, highlights that 'there is a significant global adaptation gap in health, as efforts fall far below the level required to minimize negative health outcomes' (UNEP 2018). Currently, funding allocated to adaptation projects specifically addressing health represents a tiny fraction (<0.5%) of the already limited pool of adaptation funding available (UNEP 2018).

Increasing investments to tackle present and future health challenges induced by climate change and variability is of the utmost importance to mitigate the impact on the delivery of health services. We discuss the funding landscape (global and regional funds, private-sector financing arrangements) below.

In areas where transportation infrastructure networks lack the necessary adaptation to cope with a shifting climate, the more frequent and severe climate-related events can have detrimental effects on the resilience of materials, structural integrity, maintenance schedules, and overall functionality of transportation assets and networks (Elysee 2023).

14.3 Financing the Climate-Health Nexus

To finance the protection of health from the impacts of climate change, it is essential to adopt a comprehensive approach that begins with core investments in the health sector. These investments should focus on crucial areas such as ensuring an adequate number of trained health personnel and establishing basic health infrastructure and services, which can also help address climate change risks (Corvalan et al. 2020; WHO n.d.).

Mobilizing and allocating resources effectively entails tracking funding flows and establishing baseline measurements to better coordinate the utilization of scarce public resources. It is important to conduct assessments that identify the requirements, available finance, funding gaps, and opportunities for acquiring financing from national, regional, and global sources (Romanello et al. 2022).

At the international level, there are key multilateral funding mechanisms established under the UNFCCC and the Kyoto Protocol. These mechanisms aim to facilitate financial assistance from more developed countries to those that are considered more vulnerable and have limited resources to adapt to climate change. The Global Environmental Facility (GEF) has been designated by the UNFCCC to manage the Special Climate Change Fund and the Least Developed Countries Fund. The Adaptation Fund, established under the Kyoto Protocol, and the Green Climate Fund, established more recently, are also important international funding mechanisms for climate change adaptation. Additionally, bilateral and regional channels also provide funding opportunities alongside the main international climate change-specific mechanisms. The big shift that has happened in the last 2 years particularly around President Macron's Summit for a New Global Financial Pact in Paris was to move away from purely mitigation strategies to focus on adaptation and resilience for sustainable development (Élysée 2023).

How can the Global North fund initiatives in low- and middle-income countries that are to finance the climate-health nexus at an adequate scale? Perhaps one analogy for the health sector can come from taking a leaf out of financing infrastructure projects. Under this framework, there is a need to begin by analysing different kinds of health infrastructure. Two scenarios for financing the gap for climate adaptation become apparent with specific reference to health infrastructure—those projects that are associated with cash flow streams in such a way as to enable private sector capital mobilization and those that must remain predominantly funded by the state.

Unlike in certain other infrastructure sectors, core healthcare provision will remain in most countries funded by and governed by state actors. Equally, many healthcare projects in such environments do not have any cash flows associated with them that could allow for their consideration as projects are partially or fully financeable through the private sector. The state must primarily be responsible for this category of expenditure whether on programs or infrastructure.

However, fiscal resources in low-income countries remain often extremely constrained even concerning traditional healthcare delivery, leaving aside the increased expenditures required to adapt to climate impacts.

It is in this context that some recent progress on climate finance has been made during and since COP27 and through the Summit for a New Global Financial Pact in Paris in June 2023. Launched in the context of COP27, the Bridgetown Initiative proposed by Mia Mottley, the Prime Minister of Barbados, seeks to radically transform how climate-related finance is made available to developing countries. This initiative has picked up momentum in Paris (Masterson 2023) and spawned related initiatives and ideas about closing the financing gap.

The Bridgetown Initiative proposes a comprehensive reform of development finance, specifically addressing how wealthier nations support poorer nations in coping with and adapting to climate change. It outlines three key steps. The first step involves modifying the terms surrounding funding loans and repayments. The goal is to prevent developing nations from falling into a cycle of debt crisis when their borrowing increases due to recurring disasters such as floods, droughts, and storms. Secondly, the initiative calls for development banks to provide an additional USD 1 trillion in loans to developing nations specifically for climate resilience. This includes offering discounted lending aimed at enhancing climate resilience in countries vulnerable to climate change impacts. The third step in the Bridgetown Initiative entails establishing a new mechanism, supported by the private sector, to fund climate mitigation and reconstruction efforts following climate-related disasters. It proposes the creation of a Global Climate Mitigation Trust, which has the potential to leverage USD 3–4 trillion in private funding (Persaud 2022). The Bridgetown Initiative argues that development banks providing loans and grants to developing countries to fund social and economic development are outdated. The initiative proposes to reform the Bretton Woods institutional architecture to release the constraints on such institutions so that they are allowed to leverage their balance sheets much more, releasing a huge amount of money to fund relevant initiatives.

Aside from the Bridgetown Initiative, a variety of climate finance tracks are seeking to unlock MDB finances for middle-income and low-income countries through both official and civil society coalitions as well as to focus on critical issues such as foreign currency hedging.

The emphasis of such initiatives often involves a shift from the provision of traditional credit by MDBs and IFIs to the provision of truly concessional and blended finance. (These efforts must also be seen in the context of the efforts of the G7 to cooperate in the launch of several Just Energy Transition Partnerships (JETPs) in important emerging countries; while these latter initiatives mainly focus on the energy and related sectors, some, such as the South Africa JETP, make provision for social subsidies, which may have relevance for the healthcare. Efforts have also been made to consider how 'Guarantee Funds' potentially subsidized by Northern Governments can lower real and perceived risk for the domestic savings pool, often controlled by domestic governments, which can then be directed into relevant domestic investments.

Despite all these initiatives, public finance, whether international, multilateral, or domestic, will be insufficient in meeting the financing needs of climate-sensitive and adaptive healthcare provision in emerging countries. Hence, there is an increasing focus on whether the much larger pools of private capital in the global economy can be harnessed for this and other sectors. Given the relative scale of capital markets, it is also understood that much of this private capital is likely to come from international not from domestic sources.

Much of this emphasis on private finance has evolved to a focus on 'blended finance': understanding that, very often, perceived or actual risks are seen as too great to attract significant flows of private sector finance, there is a realization that public or philanthropic capital may be most powerfully leveraged by using it to 'de-risk' projects so that they meet the risk-return needs of private capital providers. Whereas these themes apply generally to climate finance, they may also have important possibilities in the financing of health infrastructure in developing countries.

To assess the role of private capital flows—whether 'naked' or 'blended'—it is important to create a comprehensive categorization of types of private sector investors who may consider such investments. In principle this categorization would include, among many other categories:

- Risk-taking private equity and venture capital style investors willing to invest in healthcare innovation, entrepreneurship, and service delivery enterprises for a higher return, whether individually or through dedicated emerging market (EM) funds.
- Medium risk-taking infrastructure style investors seeking steady but appropriate risk-adjusted 'equity yield' style returns whether individually (e.g., strategic, sophisticated sovereign wealth funds, etc.) or through dedicated EM infrastructure style funds.
- Lower risk-seeking debt and credit investors seeking debt style returns, often either mezzanine (slightly higher risk) or senior credit (lower risk) providers, again sophisticated enough to understand and price EM project risk. These can include regulated banks, non-banking financial services companies, and the informal private credit sector.
- Those who take primary assets generated by the first three categories and invest in them on a secondary basis or securitize them.

Risk is often assessed not only to promoter quality, or the cash flow predictability of an individual healthcare project, but also to the credit quality of the EM in which the project is located as well as the predictability of the likely future performance of its currency against the dollar (or whether related hedging instruments exist and are affordable).

In addition, the level of risk appetite is quite imperative to distinguish between investors relative to the length/maturity of desired investment timelines, the size of an individual deal, and the nature of future cash flow profiles.

What is the consequence of these distinctions to the likelihood that private finance will play a crucial role in EM healthcare financing? Key elements of answering this question include:

- Country development stage: Middle income emerging economies may be seen as creating greater affordability and therefore lower risk than low-income economies.
- EMs with investment grade credit ratings will attract institutional investors to a much larger extent than those that are sub-investment grade.
- EMs whose healthcare sector has been open to private healthcare delivery for a longer period and therefore has more sophisticated private sector enterprises with a longer track record, whether at the asset level or as operators, will find it easier to attract capital more easily.
- EM private sector institutional investors need a certain critical scale to justify team time allocation and transaction costs, whereas only a few EMs have a reasonable number of such projects—those who do will find it easier to attract investment.
- EM healthcare projects that are led by senior management teams that have track records of previous success and can bridge business culture gaps will find it easier to raise private capital.

For those countries that do not have these factors fully but only in part, de-risking processes can make a significant difference in allowing private capital to flow.

Financial de-risking must play an important role in this approach in many instances. As a result of risk-return gaps, private sector potential projects may indeed require the blended finance approach. It is evident that some kinds of health infrastructure already have cashflows associated with them (like tertiary hospitals). In those cases, the need is to get bridging finance to lower the initial risk of investing in such projects. Once the investment has started, however, a benign cycle of cash flow generation may make projects sufficiently sustainable, profitable, and stable for later stages of finance not to need further blending.

It is important to recognize, however, that 'blended finance' alone is not enough to solve this problem since the available pool of subordinated or concessional finance of this type whether from Northern governments or MDBs is very unlikely to be enough to bridge the gap without more comprehensive approaches. Such integrated de-risking processes would systemically involve processes at various levels:

- Vision setting: governments must set clear, tangible, and time-bound national ambitions for health sector mitigation and adaptation objectives that create clear and predictable 'market signals'.
- De-risking of the 'real economy': such de-risking can take place at two levels. Firstly, through the convening of industrial and service supply chains in the health sector and aggregated providers of demand signals for healthcare products and services to create integrated and timebound rollout plans. Secondly, through the establishment of world-class contracting and regulatory standards.
- Support for pipeline development: Concerning supportive institutions, funding, process facilitation, etc. Also, through the creation of aggregation vehicles where market opportunities are too small or too fragmented for individual investment but can, taken together, create attractive market opportunities.
- Financial de-risking: including adopting appropriate and targeted blended finance mechanisms.
- Syndication: systematic syndication of the right types of health sector projects to the right kinds of investors, given their appetite for risk, scale, etc.

De-risking processes of this kind are most effectively conceived and implemented by engaging ab initio all relevant stakeholders

- Governments, healthcare product, and delivery providers and investors, as well as civil society actors—to work together through the stages outlined above to identify challenges that are often specific to individual countries and sub-segments, and jointly de-bottleneck and de-risk these.

It is important to recognize that these processes may also require significant institutional innovation. For example, much healthcare delivery in low-income countries is conducted through the informal economy. Whereas institutional investors may not be able to invest in these delivery providers, they may indeed be able to channel significant funds if microfinance or non-banking finance and leasing companies are established and rolled out to provide market access, service delivery, quality control, and financial discipline at the required aggregated scale.

Through integrated de-risking of this kind, limited pools of blended finance can be leveraged to maximum effect in mobilizing private sector finance.

The key to whether such de-risked and blended finance approaches may work in mobilizing private sector capital, if domestic governments have opened the sector to such investment, is fundamentally dependent on whether the project generates predictable cash flows. Not all types of healthcare, whether programs, products, services, or infrastructure, offer such cash flows. Once the framework of analysis of the type of healthcare infrastructure is completed (i.e. is it healthcare that generates predictable cash flows (private sector potential projects) or is it healthcare that comes without such cash flows (public or philanthropic potential projects), then the application of the limited amount of subsidy money that's been made available, both domestically by national governments and internationally through the COP process and the Bridgetown initiative to support developing countries in evolving their resilience infrastructure strategies can be channelled as may be appropriate.

Despite the potential of de-risking processes and blended finance approaches, we must recognize that many projects especially in low-income countries will not be private sector potential projects and therefore require sustained political commitment to finance through public institutions both domestically and by the international community. However helpful private capital may be, when properly mobilized, it cannot take the central role of government as the primary provider of climate-resilient and climate-impact-responsive health delivery.

14.4 Conclusion

The scale of climate finance required to bridge the needs of middle- and low-income emerging countries is enormous. To meet the integrated healthcare needs of the ever-increasing mass of those likely to be impacted by climate change in the decades to come, every possible source of finance needs to be mobilized:

- Domestic savings pools need to be directed into ever more effective climate health delivery systems.
- Multilateral and bilateral flows have to be scaled up enormously through sustained political will galvanized through initiatives such as the Bridgetown Initiative.
- Such flows need to move beyond the provision of credit to making available large-scale truly concessional finance as well as blended finance.
- Concessional finance should be targeted to scale the capacity of domestic governments urgently to put in place public initiatives to adapt health infrastructure and programs with sufficient scale and urgency.
- Recognizing that public and fiscal resources need to be scaled up through the mobilization of private finance, such Northern government flows should also

adopt blended finance approaches that close the real and perceived risk-return gaps for private investors in a way that is targeted to public objectives.
- Such blended finance approaches should be located within comprehensive de-risking strategies.
- Despite private finance mobilization, domestic governments must be supported to be the primary planners and implementors of the large-scale mobilization and delivery strategies required to deal with ever-increasing climate health impacts.

The primary goal of constructing healthcare facilities that are climate-resilient and environmentally sustainable is twofold: (a) to bolster their ability to safeguard and enhance the wellbeing of the communities they serve in the face of unpredictable and changing climate conditions and (b) to equip them with the means to optimize resource utilization while minimizing the release of pollutants and waste into the environment. Such healthcare institutions play a vital role in ensuring high-quality care, accessibility of services, and improved affordability by reducing operational costs. They are an integral component of achieving universal health coverage. Addressing these objectives necessitates action in four key areas essential for delivering safe and top-notch healthcare: ensuring an adequate and skilled workforce with favourable working conditions, empowering them to respond effectively to environmental challenges; implementing sustainable and secure practices for water, sanitation, and healthcare waste management; establishing sustainable energy solutions; and adopting suitable infrastructure and technologies that enable efficient healthcare facility operations. Crucially, these efforts contribute to promoting ongoing strengthening and efficiency of healthcare facilities, ultimately enhancing health outcomes and reducing disparities and vulnerabilities in their local contexts. To achieve this, we propose a comprehensive framework to address these challenges (Corvalan et al. 2020).

Evaluating the resilience of healthcare systems is crucial for policymakers as it assists in strategic planning for sustainable recovery and enables the enhancement of systems to effectively anticipate and manage both present and forthcoming crises (Haldane et al. 2021). The connection between deforestation, human-induced disruptions, and the emergence and dissemination of infections is established through various interconnected factors. It is imperative to consider the intricate interplay between proposed development initiatives and the prevalence of diseases in the Amazon region (Ellwanger et al. 2020).

Wildfires are linked to heightened rates of illness and death, yet there are significant gaps in our understanding of their impact on health. On an individual level, there is limited capacity to mitigate the adverse health outcomes resulting from wildfire exposure, emphasizing the need for collective societal efforts. In the absence of prompt measures to curb the rise in global temperatures, the interaction between wildfires and climate change is poised to create a self-reinforcing cycle, amplifying the severity of wildfires and their associated health repercussions.

Climate finance strategies do not exist in a vacuum. At the core, all such strategies will be effective only when they are part of comprehensive HNAPs, coordinated to achieve large-scale and timebound tangible results in partnership with all relevant actors from governments (domestic and international), multilateral and bilateral development banks, healthcare sector stakeholders, and the finance and investment sector, domestic and international. Such a coordinated multistakeholder approach, if motivated by urgency and at scale planning, may seek once and for all to cross a financing divide that may have seemed, given massive likely climate health impacts, hitherto too dauntingly to bridge.

References

Buchner B, Naran B, Fernandes P, Padmanabhi R, Rosane P, Solomon M, Stout S, Strinati C, Tolentino R, Wakaba G, Zhu Y, Meattle C, Guzmán S. Global landscape of climate finance 2021. Climate Policy Initiative; 2021. p. 1–48. https://www.climatepolicyinitiative.org/wp-content/uploads/2021/10/Full-report-Global-Landscape-of-Climate-Finance-2021.pdf

Cissé G, Mcleman R, Adams H, Aldunce P, Bowen K, Campbell-Lendrum D, Clayton S, Ebi KL, Hess J, Huang C, Liu Q, Mcgregor G, Semenza J, Tirado MC, Pörtner H-O, Roberts DC, Tignor M, Poloczanska ES, Mintenbeck K, et al. Health, wellbeing, and the changing structure of communities. In: Pörtner H-O, Roberts DC, Tignor M, Poloczanska ES, Mintenbeck K, Alegría A, Craig M, Langsdorf S, Löschke S, Möller V, Okem A, Rama B, editors. Climate change 2022: impacts, adaptation and vulnerability. Contribution of working group II to the sixth assessment report of the intergovernmental panel on climate change. Cambridge University Press; 2022. p. 1041–170. https://doi.org/10.1017/9781009325844.009.

Corvalan C, Prats EV, Sena A, Campbell-Lendrum D, Karliner J, Risso A, Wilburn S, Slotterback S, Rathi M, Stringer R, Berry P, Edwards S, Enright P, Hayter A, Howard G, Lapitan J, Montgomery M, Prüss-Ustün A, Varangu L, Vinci S. Towards climate resilient and environmentally sustainable health care facilities. Int J Environ Res Public Health. 2020;17(23):1–18. https://doi.org/10.3390/IJERPH17238849.

Dasandi N, Graham H, Lampard P, Jankin Mikhaylov S. Engagement with health in national climate change commitments under the Paris agreement: a global mixed-methods analysis of the nationally determined contributions. The Lancet Planetary Health. 2021;5(2):e93–e101. https://doi.org/10.1016/S2542-5196(20)30302-8.

Ebi KL, Vanos J, Baldwin JW, Bell JE, Hondula DM, Errett NA, Hayes K, Reid CE, Saha S, Spector J, Berry P. Extreme weather and climate change: population health and health system implications. Annu Rev Public Health. 2021a;42:293–315. https://doi.org/10.1146/annurev-publhealth.

Ebi KL, Vanos J, Baldwin JW, Bell JE, Hondula DM, Errett NA, Hayes K, Reid CE, Saha S, Spector J, Berry P. Extreme weather and climate change: population health and health system implications. Annu Rev Public Health. 2021b;42:293–315. https://doi.org/10.1146/ANNUREV-PUBLHEALTH-012420-105026.

Elena C-L, Prats V, Kendrovski V. Estimating the cost of health adaptation. World Health Organization; 2018. p. 1–19. https://unfccc.int/sites/default/files/resource/CGE_wenbinar%236_presentation.pdf

Ellwanger JH, Kulmann-Leal B, Kaminski VL, Valverde-Villegas JM, Da Veiga ABG, Spilki FR, Fearnside PM, Caesar L, Giatti LL, Wallau GL, Almeida SEM, Borba MR, Da Hora VP, Chies, J. A. B. Beyond diversity loss and climate change: impacts of Amazon deforestation on infectious diseases and public health. An Acad Bras Cienc. 2020;92(1):e20191375. https://doi.org/10.1590/0001-3765202020191375.

Élysée. Summit on a new global financing pact. Élysée; 2023. https://www.elysee.fr/en/emmanuel-macron/summit-on-a-new-global-financing-pact

Haldane V, De Foo C, Abdalla SM, Jung AS, Tan M, Wu S, Chua A, Verma M, Shrestha P, Singh S, Perez T, Tan SM, Bartos M, Mabuchi S, Bonk M, McNab C, Werner GK, Panjabi R, Nordström

A, Legido-Quigley H. Health systems resilience in managing the COVID-19 pandemic: lessons from 28 countries. Nat Med. 2021;27(6):964–80. https://doi.org/10.1038/s41591-021-01381-y.

Hwong AR, Wang M, Khan H, Chagwedera N, Grzenda A, Doty B, Benton T, Alpert J, Clarke D, Compton WM. Climate change and mental health research methods, gaps, and priorities: a scoping review. Lancet. 2022;6:e281–91. https://doi.org/10.1016/S2542-5196(22)00012-2.

IPCC. Summary for policymakers. In: Pörtner H-O, Roberts DC, Tignor M, Poloczanska ES, Mintenbeck K, Alegría A, Craig M, Langsdorf S, Löschke S, Möller V, Okem A, Rama B, editors. Climate change 2022: impacts, adaptation and vulnerability. Cambridge University Press; 2022. p. 1–34. https://doi.org/10.1017/9781009325844.001.

Karliner J, Slotterback S, Boyd R, Ashby B, Steele K. Health care's climate footprint: how the health sector contributes to the global climate crisis and opportunities for action I health care without harm climate-smart health care series green paper number one produced in collaboration with Arup. Health Care Without Harm and Arup. 2019.

Kinley R. Climate change after Paris: from turning point to transformation. Clim Policy. 2016;17(1):9–15. https://doi.org/10.1080/14693062.2016.1191009.

Masterson V. Barbados calls for finance reform to fight climate change. World Economic Forum; 2023. https://www.weforum.org/agenda/2023/01/barbados-bridgetown-initiative-climate-change/

OECD. Climate finance provided and mobilised by developed countries: aggregate trends updated with 2019 data. Organisation for Economic Co-Operation and Development; 2021. p. 5–21. https://doi.org/10.1787/03590FB7-EN.

Persaud, A. Bridgetown initiative calls for new global climate mitigation trust financed via special drawing rights. 2022.. https://www.brettonwoodsproject.org/2022/12/bridgetown-initiative-calls-for-new-global-climate-mitigation-trust-financed-via-sdrs/.

Pichler PP, Jaccard IS, Weisz U, Weisz H. International comparison of health care carbon footprints. Environ Res Lett. 2019;14(6):064004. https://doi.org/10.1088/1748-9326/AB19E1.

Romanello M, McGushin A, Di Napoli C, Drummond P, Hughes N, Jamart L, Kennard H, Lampard P, Solano Rodriguez B, Arnell N, Ayeb-Karlsson S, Belesova K, Cai W, Campbell-Lendrum D, Capstick S, Chambers J, Chu L, Ciampi L, Dalin C, et al. The 2021 report of the lancet countdown on health and climate change: code red for a healthy future. Lancet. 2021;398(10311):1619–62. https://doi.org/10.1016/S0140-6736(21)01787-6/ATTACHMENT/AC088142-4570-4D4D-AEB8-D760254F9A1A/MMC5.PDF.

Romanello M, Di Napoli C, Drummond P, Green C, Kennard H, Lampard P, Scamman D, Arnell N, Ayeb-Karlsson S, Ford LB, Belesova K, Bowen K, Cai W, Callaghan M, Campbell-Lendrum D, Chambers J, van Daalen KR, Dalin C, Dasandi N, et al. The 2022 report of the lancet countdown on health and climate change: health at the mercy of fossil fuels. Lancet. 2022;400(10363):1619–54. https://doi.org/10.1016/S0140-6736(22)01540-9.

UNEP. Adaptation Gap report 2018. 2018.. https://www.unep.org/resources/adaptation-gap-report-2018

UNFCCC. Report of the conference of the parties on its twenty-first session, held in Paris from 30 November to 13 December 2015. In: Addendum Part two: action taken by the conference of the parties at its twenty-first session; 2016. https://unfccc.int/documents/9097.

WHO. Climate change and health. World Health Organization; 2021. https://www.who.int/news-room/fact-sheets/detail/climate-change-and-health

WHO. Strengthening health resilience to climate change. World Health Organization; n.d. p. 1–36.

WHO, & International Institute for Sustainable Development. (2021). Health in national adaptation plans: review.. https://apps.who.int/iris/handle/10665/340915

WHO, Public Health & Environment Department (PHE), & Health Security & Environment Cluster (HSE). Operational framework for building climate resilient health systems. World Health Organization; 2015. p. 5–47.

World Bank, & International Bank for Reconstruction and Development. Economics of adaptation to climate change. 2010.. www.worldbank.org

Exploration of Interlinks Between Climate Change and Cultural Heritage: An Indian Perspective

15

R. Vani Lakshmi and Rahul Sheshan Clare

Abstract

Climate change threatens all forms of life, including the tangible and intangible elements of cultural heritage. The interplay between climate change and cultural heritage is a two-way process. While climate change causes irreplaceable losses to the epitomes of our cultural roots, tangible and intangible cultural heritage serves as a valuable resource for communities to prepare for, respond to, and recover from climate-change-induced emergencies. At the crux of the climate change, crises are the lives of the indigenous communities marred by livelihood loss and forced migration, further paving the way for complex public health challenges. This chapter revisits the latent rhythmic line linking climate change, cultural heritage, and public health through the lens of the lives of indigenous communities in India. It presents the existing challenges and opportunities for climate change research in the Indian context.

Keywords

Climate Justice · Cultural Heritage · Governance · India · Public Health · Indigenous Communities

R. Vani Lakshmi (✉)
Department of Data Science, Manipal Academy of Higher Education (MAHE), Manipal, India
e-mail: vani.lakshmi@manipal.edu

R. S. Clare
Centre for Digital Health and Technologies, Manipal Academy of Higher Education (MAHE), Manipal, India

H. Brand et al. (eds.), *Climate Change and Public Health*,
https://doi.org/10.1007/978-981-95-0895-2_15

15.1 Introduction

According to the International Panel for Climate Change (IPCC 2018), climate change refers to "a change in the state of the climate that can be identified (e.g. by using statistical tests) by changes in the mean and/or the variability of its properties and that persists for an extended period, typically decades or longer." Climate change threatens all forms of life and the tangible/intangible elements of cultural heritage. The interplay between climate change and cultural heritage is a two-way one. While climate change leads to irreplaceable losses to the epitomes of our cultural roots, the tangible/intangible cultural heritage serves as a valuable resource drawn on by communities to help them prepare for, respond to, and recover from climate-change-induced emergencies (UNESCO 2023). At the crux of the climate change crises are the lives of the indigenous communities marred by the loss of livelihood followed by forced migration, further paving the way for complex public health challenges. As the underprivileged and often ignored strata of society traverse the ocean of climate change-induced challenges, the knowledge trail derived from cultural heritage elements vanishes into its depths, depriving us of sustainable solutions passed on by oral tradition through generations. This chapter revisits the latent rhythmic line linking climate change, cultural heritage, and public health through the lens of the lives of indigenous communities in India. The chapter presents the existing challenges and opportunities for the climate change research roadmap in an Indian context.

15.2 Cultural Heritage and Climate Change

The United Nations Educational Scientific and Cultural Organization (UNESCO) defines cultural heritage as the "legacy of physical artefacts and intangible attributes of a group or society that are inherited from past generations, maintained in the present, and bestowed for the benefit of future generations" (UNESCO 2022a). Cultural heritage is a vital expression of cultural diversity and merits dedicated protection. The cultural heritage framework (Table 15.1) (Galani and Karatza 2019) enables us to understand its tangible and intangible dimensions (Galani and Karatza 2019; Martorell Carreño 2003; Mathioudakis et al. 2022). It would be futile to draw a crystal-clear line of separation between the tangible and intangible aspects of cultural heritage, primarily owing to the undeniable interlinkages between the two dimensions. The climate is central to the environmental factors of intangible heritage as changes in climate are consequences of our past actions, only to influence our present and reshape our future. Climate change impacts the remnants (tangible and intangible) of our past while re-directing our behaviours and social practices to uncharted territories of our present lifestyle.

According to UNESCO, of its 1154 World Heritage sites, "One site in five, and more than a third of natural sites, already see the threats posed by climate change as a reality"(AFP 2022). To address this, it has been at the forefront of protecting the tangible and intangible aspects of cultural heritage. India has been one of the

Table 15.1 Dimensions of cultural heritage

Themes	Tangible dimensions	Intangible dimensions
Arts and artefacts	Monuments and architectural sites, sculpture, or remnants of any form of written record, paintings, handicrafts	Language, performing arts (music, dance, theatre, and oral traditions), clothing and cuisine, folklore
Environment	Places of scenic interest (physical or biological formations, habitats of threatened species of animals and plants, landscape or natural sites that are important from the point of view of science, conservation or natural beauty (i.e. natural heritage)	Climate
People	Understanding of historic people and their lifestyle/ behaviour through tangible aspects	Behaviour of people, religion, oral traditions and expressions such as epics, fairy tales, stories, arts, social practices, customs/rituals and celebrations, cuisines, events, knowledge, and practices related to nature and the universe, traditional medicine, folk medicine, traditional handcrafts, as well as personal experiences related to significant historical events or cultural activities that shaped the historical and local identity

Adapted from (Galani and Karatza 2019).

earliest signatories to the UNESCO World Heritage Convention (UNESCO n.d.-b) and is a host to 40 sites in the World Heritage List as of 2022 (UNESCO n.d.-c). This is the sixth-highest number of sites to be featured in the list. In a way, this quantifies the climate change-induced risks to India in the context of cultural heritage.

Furthermore, UNESCO's list of Intangible Cultural Heritage and Humanity (UNESCO n.d.-a) inscribes 14 elements, including *Sankirtana* (the tradition of Vedic chanting), *Mudiyettu*, Durga Puja, Kumbh Mela, Yoga, and Novruz, highlighting the need to preserve the unique cultural experiences from the country. UNESCO is also at the forefront of protecting underwater culture heritage, forest heritage, and protection of monitoring and safeguarding of heritage in the face of global challenges.

While climate is an intrinsic component of cultural heritage, as depicted in Table 15.1, it would be unwise to overlook how culture can serve as a mechanism to minimize the impact of climate change and promote societal wellness.

UNESCO is committed to ensuring that culture is fully integrated into climate action and strategies—both as a shared global asset that must be safeguarded from the impacts of climate change and as a transversal tool that can support climate change mitigation, adaptation, and resilience. It has emphasized the role of culture in the seventeen Sustainable Development Goals (SDGs) of the 2030 agenda (Table 15.2) (UNESCO 2022b).

Table 15.2 Culture and SDGs

SDG	Role of culture in SDGs
1	Helps eradicate socio-economic poverty
2	Promotes food security through indigenous knowledge
3	Strengthens disease prevention via traditional knowledge
4	Enhances global citizenship through arts and language education
5	Advances gender equality through cultural participation
6	Protects ecosystems via heritage and community engagement
7	Encourages efficient energy use through cultural practices
8	Creates local job opportunities in creative industries
9	Drives innovation through cultural infrastructure
10	Fosters inclusion through respect for cultural diversity
11	Supports sustainable cities through creativity and heritage
12	Inspires sustainable consumption and production behaviours
13	Boosts climate resilience via traditional knowledge
14	Promotes ocean health by protecting marine heritage
15	Links biodiversity with cultural diversity for sustainability
16	Upholds human rights through respect for cultural diversity
17	Enables partnerships via artists, professionals, and policymakers

Adapted from (Climate Heritage Network 2022).

Notably, the utility of the traditional knowledge for climate resilience (SDG 13) and health (SDG 3) is unparalleled. The knowledge systems and environmental management practices of indigenous and local people provide insights enabling better management of ecological challenges, preventing biodiversity loss, reducing land degradation, and mitigating the effects of climate change. Culture, particularly traditional knowledge, also contributes to resilience and recovery in the case of natural disasters (Target 13.1 of SDG 13).

"The Congregation of Parties (COP27) (Climate Heritage Network 2022), Egypt, acknowledged the critical linkages between cultural heritage and climate change. The primary takeaways were:

- Mainstream art, culture, and heritage perspectives into climate change policymaking.
- Support climate action by cultural voices at all levels and drive engagement by wider audiences.
- Build capacity within cultural actors for durable, ambitious climate action.
- Grow the Climate Heritage Network to develop more effective international culture-climate collaboration."

15.3 Cultural Heritage and Wellness

"Wellness is not merely the absence of disease. It represents the dynamic and conscious development of the whole self. Wellness encompasses eight mutually interdependent dimensions: physical, intellectual, emotional, social, spiritual, vocational, financial, and environmental" (Stoewen 2017).

Cultural aspects interlink the social, spiritual, vocational, and environmental dimensions of wellness. The public health literature suggests three frames for public health concerns related to history and heritage: cognitive health, physiological health, and social concerns (Orthel 2017).

Cognitive health focuses on the thinking processes and knowledge people use in maintaining ontological security. Physiological conditions encompass concerns about disease, physical activity, environmental hazards, and access to life necessities, including health, food, and water. Traditional medicine is an amalgamation of indigenous knowledge/practices and public health (Prakash et al. 2014). Social concerns involve how people interact, influence others, and frame social behaviour. Re-looking at the historical lifestyles (focusing on oral traditions, folklore, and art) and social practices also provide insights to address social, cognitive, and physiological health (Fig. 15.1).

Figure 15.2 presents the interlinkage between public health and cultural heritage based on a systematic review conducted in 2022 (Orthel 2021). Several studies (Ander et al. 2013; Sofaer et al. 2021) emphasize the role of climate change in improving hedonic (subjective) and eudaimonic (psychological) well-being. Thus, culture/cultural heritage serves both as a mediator and moderator to public health and climate change.

Fig. 15.1 Culture and public health. (Orthel 2021)

COGNITIVE HEALTH	*Wellbeing* *Possession studies* *Socialization*
PHYSIOLOGICAL HEALTH	*Heritage and diet* *Exercise and the built environment* *Health concerns*
SOCIAL ISSUES	*Public welfare and good* *Social justice* *Public policy* *Culture*

Fig. 15.2 Indigenous art and culture: Some Installations from South India. (By ©Venkatesh Palimar (Palimar n.d.-a, n.d.-b))

15.4 Cultural Heritage, Climate Change, Public Health, and Adivasis: Drawing the Rhythmic Line

> Great art springs up when the artist is deeply moved from within. Even abstract art is not completely alien to Indian thought or our art, literature, and philosophy. Our religious art is full of symbols that are nothing but geometrical abstract shapes. Our villagers decorate the mud walls of their dwelling places with fantastic shapes and colours, full of symbolic significance and decorative charm. For tribal people, art is an essential part of life. This is the same all over the world. It is admitted that even modern Western abstract painters have been inspired by Hindu philosophy (Hebbar 1969). ~ K K Hebbar.

The term *Adivasi* translates to "earliest inhabitants" and refers to the earliest inhabitants of the sub-continent. The term *Adivasi* derives from the Sanskrit/Hindi word *adi*, which means of earliest times or from the beginning, and *vasi*, meaning inhabitant or resident, and it was coined by renowned political activist Thakkar Bapa in the 1930s, largely a consequence of a political movement to forge a sense of identity among the various indigenous or autochthonous people of India.

This was perhaps the earliest effort to bring together the disintegration in the indigenous community triggered by the policies of the colonial era to exert control over the length and breadth of the country systematically. The term broadly translates to the tribal population in present-day India, which accounts for 8.6% of our population based on Census 2011 data (Press Information Bureau and Ministry of Tribal Affairs 2022).

Socio-economic development of the tribal community has been one of the focus areas of independent India. However, this overlooked the dependence of tribal populations on nature and the interconnectedness between cultural heritage, health, and livelihoods. Most tribes have reservations (Survival International 2016) about stepping out of their well-defined territories as they build an earthly cocoon that protects their health and social lives while providing them with means to sustain their family.

However, over the last seven decades, *Adivasis* have continued to be at the receiving end of land alienation, loss of access and control over their homeland (Niazi 2021) (i.e. forests) and subsequent enforced displacement with poor rehabilitation/indebtedness due to a multitude of reasons, including rapid development projects and urbanization.

Indigenous women are "double-colonized" with the additional burden associated with mortality, teenage pregnancy, sexually transmitted diseases, and domestic violence (Amnesty International n.d.).

To recognize the rights of the forest-dwelling tribal communities and other traditional forest dwellers to forest resources, on which these communities were dependent for a variety of needs, including livelihood, habitation, and other socio-cultural needs, the Government of India enacted the Forest Rights Act (FRA), 2006 (Ministry of Tribal Affairs and Government of India 2023; Oxfam India 2018). However, the act is marred by implementation barriers and a lack of political will and is now tabled for amendments (as of August 2023) at the national level.

Climate change-induced environmental degradation, extreme weather events, commercialization of agriculture (Ved 2019), and socio-cultural changes have led to the systemic disintegration of *Adivasis* in the country. This has contributed to distress migration/forced migration and marginalization of *Adivasis*. For instance, research in the mental health spectrum (Singh et al. 2013; Verma et al. 2022) reveals that climate change-induced migration (Bharadwaj et al. 2021) affects not only the migrant worker but also his/her environment, encompassing all their dependents. The cost of this disintegration is also loss of livelihood, identity crisis, and the slow death of the traditional knowledge systems and folk healing traditions (Science and Society n.d.), all of which add to the public health burden of the country. For instance,

The Aka tribe in Arunachal Pradesh has its way of treating fever and cough. The trick is to crush 10 grams of fresh leaves and stems of a "Syowum" plant, strain it through a cloth, and apply it on the body. Although the medicinal plant has a scientific name, it is the native language that enables us to decode its secrets (Kalia 2021).

The language faces extinction today, as is the case of indigenous knowledge. This is a fraction of the mammoth challenges that the indigenous population faces globally. Fig. 15.3 presents a conceptual framework associated with the role of environmental change as a 'driver' of migration and its potential impacts on health (Schwerdtle et al. 2021). It is pertinent to note that climate-change-induced variabilities in the environment could increase the mobility of the population or immobilize/trap a stratum of the population, both of which could lead to increased psycho-social health risks, inadequate access to health services, and food insecurities. The striking difference is in the challenges associated with adaptation to the environment and cultural aspects at the destination sites when it comes to the migrated population. Additionally, there is an irreplaceable loss of ancestral roots (a sense of identity) and the indigenous knowledge/skills passed on by generations.

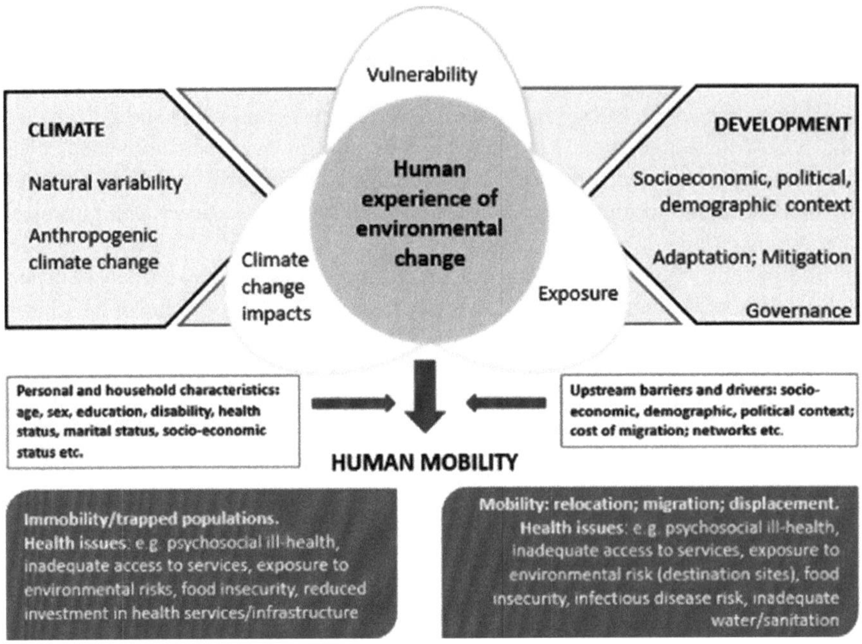

Fig. 15.3 Human experience of environmental change. (By Schwerdtle et al. (2021). International Journal of Environmental Research and Public Health)

The Indigenous communities complete the strand of the losses linked to climate change, public health, and cultural heritage. Much of the efforts at the various stakeholder levels are associated with improving the socio-economic status, protecting tangible heritage, and creating infrastructure enabling the communities to adapt to the urbanized culture, all of which fail to address the crises linked to losing one's identity. As the most affected people (Tripathi 2019) in the most affected area, this scenario represents an example of the antonym of climate justice.

15.5 Climate Justice: Global Experience in the Lens of Cultural Heritage

Climate justice (Manzo 2021) in a broader context refers to "a set of rights and obligations, which corporations, individuals and governments have towards those vulnerable people who will be in a way significantly disproportionately affected by climate change." The IPCC is at the forefront of advocating climate-resilient development to ensure climate justice across its member countries. In its AR6 Report (2023), the IPCC identified the enablers and disablers of climate-resilient development (see Fig. 15.4).

Click or tap here to enter text.

Enablers	Disablers
Inclusive governance	Poverty, inequity and injustice
Diverse knowledges and values	Economic, institutional, social and capacity barriers
Finance and innovation.	Siloed responses.
Integration across sectors and time scales.	Tradeoffs with SDGs.
Ecosystem stewardship, Synergies between climate and development actions.	Lack of finance, and barriers to finance and technology
Behavioural change supported by policy, infrastructure and socio-cultural factors	Poverty, inequity and injustice

Fig. 15.4 Enablers and disablers of climate resilient development. (Adapted from (IPCC 2023). Intergovernmental Panel on Climate Change)

It is pertinent to note that cultural factors, diverse knowledge, and values play a significant role in climate-resilient development. This calls for identifying novel pathways to revive our cultural heritage and reshape our public health landscape in the wake of climate resilience.

The initiatives taken at macro- and micro-levels to save the planet for the future are unprecedented. The International Co-Sponsored Meeting on Culture, Heritage and Climate Change served as a gamechanger for the interlinkages between climate change, public health, and cultural heritage as it brought out fresh perspectives on the role of cultural heritage as a "wealth of knowledge to contribute to grapple with climate change" (Morse 2023).

In its efforts to preserve cultural heritage, the European Commission developed a comprehensive multi-stakeholder policy dialogue that identified the participatory governance model to unravel solutions that ensure the long-term value and sustainability of cultural heritage in the European Union. The year 2018 was observed as the European Year of Cultural Heritage (European Commission and Directorate-General for Education 2018), further amplified by social media with #EuropeforCulture. Through the Erasmus+ initiative, several projects on cultural heritage focused on youth, higher education, vocational education and training, cultural tourism, sports, school, and adult education. After the success of the observance, the European Framework for Action on Cultural Heritage (European Commission 2019) was formulated to capture and scale up heritage-related activities at the European level.

"The framework outlines around 60 actions for the European Commission to implement in 2019 and 2020, focusing on five main themes. These include making cultural heritage more accessible and inclusive for European region, using heritage to promote smart, sustainable, and cohesive development, and protecting endangered cultural assets to build a more resilient Europe. It also emphasizes the importance of driving innovation through cultural research and knowledge-sharing, as well as strengthening global partnerships through international cultural cooperation."

Such initiatives provide a mechanism to promote cultural heritage and reinforce a sense of belonging while enabling them to re-discover and engage with the Government. UNESCO (UNESCO n.d.-d) has also established the "International Fund for Cultural Diversity," a multi-donor fund per the 2005 Convention on the Protection and Promotion of the Diversity of Cultural Expressions, to support multiple projects in developing countries annually. British Council (British Council n.d.) also promotes cultural heritage through funded projects in developing countries, particularly in the Middle East. Another exciting initiative in preserving the non-tangible cultural heritage is "The Living Tongues Project" (Living Tongues Institute for Endangered Languages n.d.). It is a US-based not-for-profit initiative that scientifically documents endangered languages and assists communities with maintaining and revitalizing knowledge of their native languages. Such active engagements on the global front have paved the way for a renewed interest in considering cultural heritage as one of the crucial paths to reducing the impact of climate change on public health.

15.6 Cultural Heritage as the 'Light at the End of the Tunnel'

India is a land of unparalleled cultural diversity and historical significance. Its cultural heritage is a long-standing testament to the traditions, beliefs, and values of its people. On the national front, India has been the frontrunner in reviving the lost roots of climate resilience. As early as 2008 (IndBiz et al. 2023), the country globalized, fostering cultural exchange with indigenous communities and bolstering diplomatic ties and trade (Fig. 15.5).

With India's rise to the G20 Presidency and the 18th G20 New Delhi Summit under the theme of "*Vasudeva Kutumbakam* – One Earth, One Family, One Future" drawn from the Maha Upanishad, it is pertinent to note that the country has placed immense focus on cultural heritage and its role in sustainable and climate-resilient development. The G20 Leader's Declaration ("G20 New Delhi Leaders' Declaration," 2023) highlights the role of culture in the following manner:

"We urge full recognition of culture as a vital force for achieving the SDGs and support its inclusion as a standalone goal post-2030. We commit to combating illicit trafficking of cultural property and call for its restitution. We also advocate protecting living cultural heritage and intellectual property from misuse, ensuring sustainability for communities and Indigenous people."

India's perspective on the intersection of climate change, public health, and cultural heritage highlights the critical need to address these interconnected issues for a sustainable future. Climate change impacts public health through several manifestations, necessitating comprehensive public health responses that integrate climate resilience and indigenous knowledge derived from cultural heritage.

Fig. 15.5 "The Earth" installation at MAHE, Manipal. (By © Chitranjan Moharana (Manipal Matters 2018; Moharana n.d.))

Preserving cultural heritage, harnessing art and culture for climate change awareness, and promoting sustainable livelihoods are crucial components of India's response to climate change. India can pave the way for a sustainable and inclusive future by building resilience and integrating climate change into various sectors. In the last decade, the preservation of living (i.e. intangible) cultural heritage has been a priority for the governments at the national and state levels. Upliftment of indigenous communities through increased access to healthcare (through government schemes/policies, mobile hospitals, better healthcare), education (scholarships, financial support), and basic infrastructure (food, shelter and clothing, electricity, water, and related resources) for sustainable living have been some of the mechanisms employed by the Government. Heritage museums and identification of the otherwise lesser-known tourism clusters enabling better visibility to the treasure trove of knowledge and skills among the indigenous community have been one of the focus areas for the Government. The Make in India and the Start-up India campaign focused on small and medium enterprises borne out of the skill sets of the indigenous community (Tribal Co-Operative Marketing Development Federation of India Limited et al. n.d.) have also garnered attention globally. Since 2017, the Ministry of Tribal Affairs has observed the "Aadi Mahotsav" to promote the traditional art and culture of the tribal population on the national front (NITI Aayog 2022; Press Information Bureau et al. 2021). Furthermore, the last half a decade has also been a witness to awards and recognitions focusing on the contributions of the tribal community towards environment and climate change. Additionally, over 66 unique tribal products received the Geographical Indications (GI) tag in the year 2021 (PIB Delhi 2021), emphasizing the community's rights to their intellectual property.

For instance, over a year ago, Tulasi Gowda, a 72-year-old tribal woman belonging to the Halakki tribe from Karnataka, was conferred the Padma Shri for her contribution to protecting the environment. She is known as the "Encyclopedia of the Forest" (India Today Web Desk 2021) owing to her vast knowledge of diverse species of plants and herbs, and this national recognition is indeed an honour to the traditional knowledge systems. At the national and state levels, several ministries/departments, including tribal affairs, education, women and child development, health, and family welfare, are functioning in cohesion to include culture as a quintessential component in climate-resilient and sustainable development.

India currently ranks eighth among the 59 countries and the European Union in the recently released annual Climate Change Performance Index (Burck et al. 2022) (CCPI), an independent monitoring tool for tracking climate protection performance, enhancing transparency in international climate politics. This re-affirms the country's commitment to strive towards climate resilience and projects the success of the initiatives undertaken at the national and state levels to integrate culture into the climate change resilience frameworks.

Another feather in the cap of India's journey to culture-inclusivity in a sustainable business model is centred on what is known as experiential tourism (Khokhar 2023). Also known as immersion travel, experiential tourism allows individuals to experience tangible and intangible aspects of cultural heritage through active and meaningful engagements with the place of visit. The country has witnessed a significant rise in the use of the private-public partnership model, engaging artists and the local community in a manner that preserves heritage and traditional knowledge, overcoming the burden of development-induced displacement and associated stress factors. While this model enhances community involvement and empowerment, there is a need for more robust legal and policy support to ensure transparency and sustainability to the business while ensuring climate resilience and minimal environmental impact (Figs. 15.6 and 15.7).

Technological advancements also open doors to infinite possibilities to integrate the invaluable traditional ecological knowledge that can contribute to climate change adaptation and mitigation and novel integration mechanisms in the renewable energy sector. By incorporating indigenous practices into modern conservation efforts, India can strengthen its climate-resilient sustainable development efforts, offering positive health outcomes.

There is also a growing need to significantly enhance the awareness of traditional knowledge hidden in several smaller clusters within the country and educate the indigenous community about the economic aspects and intellectual property rights associated with cultural heritage. In the past, innovations involving traditional knowledge have often been embroiled in biopiracy (Chakrabarty and Kaur 2021) issues and enhanced education and awareness are the first corrective action. With increased access to mobile devices and web services, simulation and gaming

Fig. 15.6 A glimpse of experiential tourism in Karnataka. (By ©Venkatesh Palimar (Palimar n.d.-a)).

Fig. 15.7 A glimpse of experiential tourism in India. (By ©Venkatesh Palimar, ©Chitaranjan Moharana (Palimar and Moharana n.d.))

approaches offer an exciting pathway to improve awareness and educate the citizens. While there have been several references to Indian culture (Bhavani 2022) in games, including Lara Croft: Tomb Raider, Uncharted: The Lost Legacy and Raji: An Ancient Epic, there is still a scope to use simulation and gaming as a means to provide an engaging deep-dive into the awareness aspects of tangible and intangible cultural heritage, enhancing cultural heritage literacy for public health and climate resilience at the national and global front.

With India's rise in G20 and several similar platforms in recent years, international cooperation is needed to identify innovative mechanisms that interlink climate resilience, cultural heritage, and public health through international collaborations and partnerships with international organizations. Thus, there is a need to have a more extensive research picture that perceives cultural heritage as the light at the end of the tunnel for solutions to address climate resilience. Multi-stakeholder engagements involving academia and research entities must strive towards integrating art and humanities, science and technology, and commerce to arrive at innovative solutions to overcome the challenges witnessed at the intersection of climate change, public health, and cultural heritage. This will ensure climate justice and minimize societal inequities at large.

> Humanity can address a lot of the suffering that occurs in the world and make things a lot better. I think a lot of times people are quite sort of negative about the present and about the future, but really, if you are a student of history, when else would you really want to be alive? ~ Elon Musk (Clifford 2019).

This is perhaps the best time for multi-stakeholder initiatives to address the challenges of climate change, public health, and cultural heritage.

References

AFP. Climate change 'main threat' for world heritage sites. The Hindu. 2022; https://www.thehindu.com/sci-tech/energy-and-environment/climate-change-main-threat-for-world-heritage-sites/article66153258.ece

Amnesty International. Indigenous peoples. Amnesty International; n.d.. Retrieved September 16, 2023, from https://www.amnesty.org/en/what-we-do/indigenous-peoples/

Ander E, Thomson L, Noble G, Lanceley A, Menon U, Chatterjee H. Heritage, health and well-being: assessing the impact of a heritage focused intervention on health and well-being. Int J Herit Stud. 2013;19(3):229–42. https://doi.org/10.1080/13527258.2011.651740.

Bharadwaj R, Hazra S, Reddy M, Das S, Kaur D. Connecting the dots and social protection; 2021. p. 56. https://www.iied.org/20591iied

Bhavani DK. The rise of India-made and India-centric video games pushes 'play' on representation. The Hindu. 2022; https://www.thehindu.com/sci-tech/technology/india-developing-homegrown-video-games-for-representation-in-gaming/article38407127.ece

British Council. Cultural protection fund. British Council; n.d.. Retrieved September 17, 2023, from https://www.britishcouncil.org/arts/culture-development/cultural-protection-fund

Burck J, Uhlich T, Bals C, Höhne N, Nascimento L, Tavares M, Strietzel E. Climate change performance index 2023. Germanwatch; 2022. https://ccpi.org/download/climate-change-performance-index-2023/

Chakrabarty SP, Kaur R. A primer to traditional knowledge protection in India: the road ahead. Liverpool Law Rev. 2021;42(3):401–27. https://doi.org/10.1007/S10991-021-09281-4/FIGURES/3.

Clifford C. Musk, Gates and Buffett say now is the best time to be alive. CNBC. 2019; https://www.cnbc.com/2019/07/19/musk-gates-and-buffett-say-now-is-the-best-time-to-be-alive.html

Climate Heritage Network. COP27 – Climate Heritage Network. Climate Heritage Network. 2022; https://www.climateheritage.org/cop27-home

European Commission, & Directorate-General for Education. Creative Europe: rediscovering our cultural heritage. Publications Office of the European Union; 2018. p. 1–23. https://doi.org/10.2766/179502.

European Commission, D.-G. for E. European framework for action on cultural heritage. Publications Office of the European Union; 2019. p. 1–34. https://doi.org/10.2766/949707.

G20 New Delhi Leaders' Declaration. 2023. G20.

Galani A, Karatza A. European identity and European heritage: a critical approach of Greek, British and Spanish geography textbooks. Am J Educ Res. 2019;7(12):966–75. https://doi.org/10.12691/EDUCATION-7-12-11.

Hebbar KK. Lectures on art. Prasaranga: University of Mysore; 1969.

IndBiz, Ministry of External Affairs, & Government of India. India's Twinning Agreements -s IndBiz | Economic Diplomacy Division. IndBiz; 2023. https://indbiz.gov.in/trade/indias-twinning-agreement/.

India Today Web Desk. Meet Padma Shri Tulasi Gowda, the barefoot environmentalist known as "Encyclopedia of Forest". India Today. 2021; https://www.indiatoday.in/india/story/padma-shri-tulasi-gowda-barefoot-environmentalist-encyclopedia-of-forest-1874499-2021-11-08

IPCC. Glossary — Global Warming of 1.5 °C. In: Masson-Delmotte V, Zhai P, Pörtner H-O, Roberts D, Skea J, Shukla PR, Pirani A, Moufouma-Okia W, Péan C, Pidcock R, Connors S, Matthews JBR, Chen Y, Zhou X, Gomis MI, Lonnoy E, Maycock T, Tignor M, Waterfield T, editors. Global warming of 1.5°C. An IPCC Special Report on the impacts of global warming of 1.5°C above pre-industrial levels and related global greenhouse gas emission pathways. Cambridge University Press; 2018. p. 541–62. https://doi.org/10.1017/9781009157940.008.

IPCC. AR6 Synthesis Report: Climate Change 2023. Contribution of Working Groups I, II and III to the sixth assessment report of the intergovernmental panel on climate change. In: Core Writing Team, Lee H, Romero J, editors. Climate change 2023. Intergovernmental Panel on Climate Change; 2023. p. 1–184. https://doi.org/10.59327/IPCC/AR6-9789291691647.

Kalia S. As indigenous languages Die, India loses vital means of preserving biodiversity. The Swaddle; 2021. https://theswaddle.com/as-indigenous-languages-die-india-loses-vital-means-of-preserving-biodiversity/

Khokhar R. Experiential tourism is booming in the India of many cultures. Vogue India. 2023; https://www.vogue.in/content/experiential-tourism-is-booming-in-the-india-of-many-cultures

Living Tongues Institute for Endangered Languages. Living tongues institute for endangered languages | We are a non-profit research institute dedicated to documenting endangered languages around the world. Living Tongues Institute for Endangered Languages; n.d.. Retrieved September 16, 2023, from https://livingtongues.org/

Manipal Matters. Signaling 'The Earth' Awareness. Manipal Matters Newsletter. 2018; https://manipalmatters.manipal.edu/munews/news-list/article/signaling%2D%2Dthe-earth%2D%2Dawareness.html

Manzo R. Climate Equity or Climate Justice? More than a question of terminology. IUCN, International Union for Conservation of Nature and Natural Resources; 2021. https://www.iucn.org/news/world-commission-environmental-law/202103/climate-equity-or-climate-justice-more-a-question-terminology

Martorell Carreño A. Cultural routes: Tangible and intangible dimensions of cultural heritage; 2003. p. 1–7. http://www.international.icomos.org/victoriafalls2003/papers/A1-5%20-%20Martorell.pdf

Mathioudakis G, Klironomos I, Partarakis N, Papadaki E, Volakakis K, Anifantis N, Papageorgiou I, Pavlidis SA, Antona M, Stephanidis C. InCulture: A Collaborative Platform for Intangible Cultural Heritage Narratives. Heritage. 2022;5(4):2881–903. https://doi.org/10.3390/HERITAGE5040149.

Ministry of Tribal Affairs, & Government of India. Forest Rights Act. Ministry of Tribal Affairs, Government of India; 2023. https://tribal.nic.in/FRA.aspx

Moharana C.. "The Earth" installation at MAHE, Manipal n.d..

Morse I. Cultural heritage is an essential resource for climate change science | PreventionWeb. Mongabay; 2023. https://www.preventionweb.net/news/cultural-heritage-essential-resource-climate-change-science-reports-say

Niazi S. Forest-dwelling communities in India continue to lose their lands, livelihoods. Anadolu Agency. 2021; https://www.aa.com.tr/en/asia-pacific/forest-dwelling-communities-in-india-continue-to-lose-their-lands-livelihoods/2328352

NITI Aayog. Local to Global: Promotion of Tribal Products. NITI Aayog; 2022. https://www.niti.gov.in/local-global-promotion-tribal-products

Orthel BD. Preserving Public Health: A Literature Review. In: Proceedings from the International Association of Societies of Design Research (IASDR) Conference. University of Cincinnati; 2017.

Orthel BD. Linking public health and heritage work. Int J Herit Stud. 2021;28(1):44–58. https://doi.org/10.1080/13527258.2021.1903969.

Oxfam India. Top 4 reasons why the Forest Rights Act fails. Oxfam India. 2018; https://www.oxfamindia.org/blog/forest-rights-act

Palimar, V. A glimpse of experiential tourism in Karnataka. n.d.-a.

Palimar, V. Indigenous art and culture: some installations from South India. n.d.-b.

Palimar, V., & Moharana, C. A glimpse of experiential tourism in India. n.d..

PIB Delhi. Spotlight on exquisite GI-tagged products at the ongoing Tribes India AadiMahotsav at DilliHaat. Press Information Bureau; 2021. https://pib.gov.in/PressReleasePage.aspx?PRID=1774736

Prakash N, Ansari MA, Punitha P, Sharma PK. Indigenous traditional knowledge and usage of folk bio-medicines among Rongmei Tribe of Tamenglong District of Manipur, India. Afr J Tradit Complement Altern Med. 2014;11(3):239–47. https://doi.org/10.4314/AJTCAM.V11I3.34.

Press Information Bureau, & Ministry of Tribal Affairs. Honouring and Empowering the Adivasis of India. Press Information Bureau; 2022. https://pib.gov.in/FeaturesDeatils.aspx?NoteId=151222&ModuleId%20=%202

Press Information Bureau, Government of India, & Ministry of Tribal Affairs. Manufacturing of tribal products and skill upgradation of tribal artisans under instructional support for development and marketing of tribal product/produce. Press Information Bureau; 2021. https://pib.gov.in/Pressreleaseshare.aspx?PRID=1704916

Schwerdtle PN, Baernighausen K, Karim S, Raihan TS, Selim S, Baernighausen T, Danquah I. A risk exchange: health and mobility in the context of climate and environmental change in Bangladesh-a qualitative study. Int J Environ Res Public Health. 2021;18(5):1–17. https://doi.org/10.3390/IJERPH18052629.

Science and Society. Folk healing traditions | History and science of indian systems of knowledge. Science and Society. n.d.; Retrieved September 16, 2023, from https://www.ncbs.res.in/HistoryScienceSociety/content/folk-healing-traditions

Singh PK, Singh RK, Biswas A, Rao VR. High rate of suicide attempt and associated psychological traits in an isolated tribal population of North-East India. J Affect Disord. 2013;151(2):673–8. https://doi.org/10.1016/J.JAD.2013.07.018.

Sofaer J, Davenport B, Sørensen MLS, Gallou E, Uzzell D. Heritage sites, value and wellbeing: learning from the COVID-19 pandemic in England. Int J Herit Stud. 2021;27(11):1117–32. https://doi.org/10.1080/13527258.2021.1955729.

Stoewen DL. Dimensions of wellness: Change your habits, change your life. Can Vet J. 2017;58(8):861–2. /pmc/articles/PMC5508938/

Survival International. India: Forest tribe "will die out" if evicted from ancestral land. Survival International; 2016. https://www.survivalinternational.org/news/11254

Tribal Co-Operative Marketing Development Federation of India Limited, Ministry of Tribal Affairs, & Govt. of India. Product categories | TRIFED. Tribes India. n.d.; Retrieved September 16, 2023, from https://trifed.tribal.gov.in/retail-market/product-category

Tripathi B. India 5th most vulnerable to climate change fallouts, its poor the worst hit. IndiaSpend; 2019. https://www.indiaspend.com/india-5th-most-vulnerable-to-climate-change-fallouts-its-poor-the-worst-hit/

UNESCO. Tangible Cultural Heritage Expertise Committee. UNESCO Türkiye Millî Komisyonu; 2022a. https://unesco.org.tr/Pages/654/295/Tangible-Cultural-Heritage-Expertise-Committee

UNESCO. The Tracker Culture & Public Policy | MONDIACULT Special issue n°1: Culture in the 2030 Agenda | UNESCO. UNESCO; 2022b. https://www.unesco.org/en/articles/tracker-culture-public-policy-mondiacult-special-issue-ndeg1-culture-2030-agenda

UNESCO. Operational principles and modalities for safeguarding intangible cultural heritage in emergencies – intangible heritage – Culture sector. UNESCO; 2023. https://ich.unesco.org/en/operational-principles-and-modalities-in-emergencies-01143

UNESCO. Browse the lists of intangible cultural heritage and the register of good safeguarding practices – intangible heritage – Culture sector. UNESCO; n.d.-a. Retrieved September 16, 2023, from https://ich.unesco.org/en/lists?multinational=3&display1=countryIDs#India

UNESCO. UNESCO world heritage centre – convention concerning the protection of the world cultural and natural heritage. n.d.-b.. Retrieved September 16, 2023, from https://whc.unesco.org/en/conventiontext/

UNESCO. UNESCO World heritage centre – world heritage list statistics. UNESCO; n.d.-c. Retrieved September 16, 2023, from https://whc.unesco.org/en/list/stat/#s7

UNESCO. What is the IFCD? | Diversity of cultural expressions. UNESCO; n.d.-d. Retrieved September 17, 2023, from https://www.unesco.org/creativity/en/ifcd/what-ifcd

Ved S. Indigenous Indian produce is dying, and this is how we're to blame. Vogue. 2019; https://www.vogue.in/culture-and-living/content/indigenous-indian-local-produce-is-dying-and-this-is-how-were-to-blame

Verma P, Sahoo KC, Mahapatra P, Kaur H, Pati S. A systematic review of community-based studies on mental health issues among tribal populations in India. Indian J Med Res. 2022;156(2):291–8. https://doi.org/10.4103/IJMR.IJMR_3206_21.

9789819508945